Introduction

This manual is designed to enable auto repair shops, dealers and skilled individuals to convert full-size pickup trucks to Plug-in Electric Hybrid Vehicles and to reduce their fuel consumption by 50 % while driving on electric energy. The objective is to help pickup truck owners, and the U. S. as a whole, reduce petroleum consumption without sacrificing the performance and utility of these workhorse vehicles. The technology offers owners an insurance policy against rising fuel prices and the decrease in value of the vehicles themselves, which we saw during the fuel price spike of 2008.

The manual includes background information on the rationale and evolution of the approach taken and data from actual experience performing the conversions and driving the vehicles. Step by step instructions are provided for the conversion process and for start up and maintenance afterwards. Drawings and instructions are available for those who wish to make their own components, and a catalog is provided for purchased parts.

The technology has been demonstrated on the popular Ford F-150. The drawings and instructions derive from conversion of 2-wheel drive, long-wheelbase generation 11 (2004-2008) models. They are believed to be applicable to generation 10 (1998-2003) and Generation 12 (2009-2014), but accurate measurements should be made for comparison with the Gen 11 truck dimensions shown in **Drawings 9, 10 and 18**. Similarly for 4WD and short wheelbase Fords and full-size trucks of other manufacturers, accurate measurements will show whether the conversion will fit, and any modifications that are needed to mount the components. Future editions of this manual are planned to apply the technology to other vehicles. Body-on-frame SUVs and tradesman vans, such as the Ford E series, offer a promising opportunity. Conversion of still larger vehicles is a possibility.

Application to smaller pickups such as the Ford Ranger is not advisable. The original vehicle fuel consumption is less, the returns are less, and the space for the conversion is less. Application to passenger cars and crossovers is still less likely to be successful. The conversion is not applicable to front wheel drive vehicles. Unibody vehicles offer very little room underneath, and the fuel consumption is still lower than small trucks, offering less incentive for conversion. Conversion of older model body-on-frame passenger vehicles with more room underneath and higher fuel consumption is possible.

Because of the various uses which may be made of this book by individuals and organizations of varying skill levels, the author cannot assume any responsibility for the results. NO WARRANTY OF FITNESS OR PERFORMANCE, EXPRESS OR IMPLIED IS PROVIDED, AND NO RESPONSIBILITY FOR DIRECT OR CONSEQUENTIAL DAMAGES RESULTING FROM UNSATISFACTORY RESULTS CAN BE ACCEPTED. The sole responsibility assumed by the author is that the statements in the text are accurate to the best of his knowledge and represent actual experience converting and operating PHEV pickups. The drawings have been carefully checked and are believed to be accurate, but may still contain errors. Suggestions for improvement and corrections are welcome.

To order a set of drawings e-mail partnerhsips1@verizon,net

Table of Contents

List of Figures

Chapter 1

Chapter 2

Chapter 3

Chapter 5

List of Tables

Chapter 1

Chapter 1: PHEV Pickup Trucks

Rationale, Concept, Performance

The Lithium Economy

Imagine a world in which there is a magic bottle in which lightning can be stored and then let out to do whatever work needs to be done. The bottle can be refilled from any electric outlet, the efficiency of storing and retrieving the lightning is over 90% and the bottle lasts a lifetime.

Put a bottle in your vehicle and say goodbye to petroleum. The vehicle will not pollute the air. It will not emit greenhouse gases. It will be powerful, simple, reliable and virtually maintenance-free.

Put a bottle in your boat and it will not pollute the water. The motor will be silent and powerful, allowing you to race to your destination and hear the wildlife on the way.

Put a bottle in an airplane, admittedly a high efficiency airplane like a sailplane, and soar hundreds of miles silently and swiftly.

And all of this powered by indigenous U.S. sources of energy, from the existing electrical grid, requiring no innovations in infrastructure and only evolutionary changes in its growth to achieve environmentally sound, economically profitable energy independence.

That magic bottle is here![i] It is called the lithium-ion battery, and it makes converting your low-mileage pickup truck to a high-mileage, high-performing vehicle practical.

Electric Vehicles (EV)s have been around since the beginning of the automotive era a century ago. They have many advantages. Why have they not dominated? The batteries! Up to now the choice has been lead-acid batteries with a specific energy of 35 Watt hours per kilogram, and it just isn't enough. The General Motors EV-1 was an all out attempt to create a modern EV with lead-acid batteries by a thoroughly professional team, and it was brilliantly successful, except the car would only go 90 miles on a charge and it just wasn't enough. With nickel-metal hydride batteries the range was 150 miles, which was enough, but the price was outlandish. With lithium batteries at 150 Wh/kg the EV-1 would go 300 miles and be a great success, as the Tesla has shown, at a price that is still pretty outlandish, but coming down fast. The success of the Nissan Leaf and a number of other EV entrants shows that a new era in automotive energy supply is almost here enabled by the phenomenal performance of lithium-ion batteries.

What is a PHEV?

A Plug-in Hybrid Electric Vehicle (PHEV), as the name implies, is a hybrid electric vehicle that can be plugged in to the electric grid to recharge its batteries and thus derive some of its energy from electricity. Hybrid vehicles such as the Toyota Prius derive some of their propulsion <u>power</u> from electricity stored on board, but all of the <u>energy</u> comes from gasoline. The PHEV modifies this by recharging a larger battery with electric energy from the electric grid, further reducing consumption of petroleum fuel in favor of domestically produced coal, natural gas, nuclear and renewables.

Why Convert?

The primary incentive to convert any vehicle, particularly pickup trucks and vans with poor fuel economy, to a PHEV is to reduce petroleum consumption and the consequent air pollution. In the current environment this also saves money and provides insurance against future increases in the cost of fuel. It also provides insurance against a decrease in value of the vehicles themselves due to high fuel cost, which was a notable feature of the last spike in fuel prices in 2008.

Why Pickups?

Our focus on pickup trucks is due to our major objective of reducing petroleum consumption for both national security and climate change reasons. To save fuel, go where the fuel consumption is, and predominantly it is by the heavy vehicles. Month after month, decade after decade, Ford sells 50,000+ F-150s. GM sells about the same number between the Silverado and the Sierra. Dodge, Toyota and Nissan add more. The total is well over a million vehicles a year with mileage typically less than 20 mpg in the past and in the low 20's even with turbocharged engines and aluminum bodies.

Many users depend on their vehicles for their livelihood, and high fuel prices are a serious threat. Switching to a smaller, lighter more fuel efficient vehicle is not an option. If you need to haul heavy loads or get to the job site with all your tools and materials, you need a big vehicle. Properly done, a PHEV conversion will actually improve performance in terms of towing and acceleration due to the combined action of the original IC engine and the supplemental electric drive.

Yet another major reason for focusing on pickups is that they have the room underneath to allow for a supplementary drive. Most automobiles these days are as tightly crammed with components as a jet fighter, and for the same reason. They want the most useable inside space for the least outside envelope.

Problems to be Solved

The first problem is that these heavy vehicles are **heavy**. 5000 lb empty is not uncommon, and loaded they go well over 7000. They have big engines for a reason. They need them for acceptable performance. This is a problem for an electric pickup truck drive in that, unlike a light passenger car, there is a requirement for massive starting torque and sustained high power on the highway. The AC Propulsion 150 KW drive can do it, but not your average DC system.

The second problem is actually connecting to the drive train of the vehicle. The ideal would be to insert the electric motor next to or in place of the torque converter so that it could supplement the Internal Combustion (IC) engine directly and take advantage of torque multiplication in the transmission. This is a major development requiring detailed integration of the electrical machinery with the engine-transmission of each vehicle, more suitable to a new vehicle design than to a retrofit. The next best alternative is to couple to the drive shaft while still providing speed reduction from the electric motor to minimize its size and weight and permitting flexibility of the drive shaft to move and vibrate.

This brings us to the final problem, namely cost. For electric propulsion to become widespread, the cost of conversion has to be in the vicinity of $15-25,000. A $30,000 drive system with a $30,000 battery is not economically feasible for most pickup truck owners.

The Supplemental PHEV Concept

Consideration of these problems brought us to the supplemental hybrid concept. We cannot afford to reproduce the massive, over-capacity IC drive system that the original vehicle manufacturer has already provided. We can only afford to supplement that drive train in a mode that allows electricity to provide the bulk of the <u>energy</u> (for as long as the batteries can provide it) but not the bulk of the <u>power</u>. You already paid for the power when you bought the truck. What you want is access to affordable, clean, U.S.- made electric energy.

This approach has another major cost advantage in that the changes required by the conversion are minimized. Since the IC drive train is retained, and is running, it can perform all the auxiliary functions that it always has including power steering, power brakes, starter battery charging, air conditioning and heating. This saves an enormous amount of time and money that would be required to duplicate all those functions in a complete electric conversion. The entire conversion to a supplemental PHEV can be accomplished without opening the vehicle's hood.

Yet another advantage of the supplemental hybrid concept is the efficient use of battery capacity. The battery can be sized to cover the daily commute with some margin to provide for optimum depth of discharge. Extra driving is provided for seamlessly by the always-running IC engine, and the only penalty is more frequent fuelling for trips beyond the range of the battery. Having driven both pure Battery Electric Vehicles and PHEVs, the author can testify with feeling to the peace of mind with regard to battery depletion that a PHEV can provide, (while maintaining the satisfaction of taking some of the bread out of the mouth of OPEC and cleaning up the air we breathe).

Organization of this Manual

The remainder of this chapter is a description of the development work that we have done on the supplemental PHEV conversion concept and of the results achieved. Chapter 2 discusses the basic component technologies involved and what you can expect in terms of cost and savings if you decide to do a conversion. Chapter 3 contains detailed instructions on doing the conversion and Chapter 4 covers Startup and troubleshooting, Chapter 5 contains instructions on making the specialized components required if you have the shop facilities required. A set of Drawings is available to enable this to be done, and they are referred to in this manual in boldface. Chapter 6 is a catalog of sources for purchased components along with engineering data and a bibliography.

Figure 1.1. Auxiliary electric accelerator pedal mounted immediately behind the IC accelerator.

Technology Background

This section describes the evolution of a method to apply the supplemental PHEV approach to convert full size pickup trucks, (and other heavy vehicles such as vans, SUVs and even school busses), to Plug-in Hybrid Electric Vehicles. The method uses a <u>flexible and compliant</u> power transmission subsystem connecting one or more electric motors to the drive shaft of the vehicle to supplement the existing internal combustion power train. The electric drive provides as much propulsion energy as it can, while retaining the IC engine unchanged to provide all the ancillary services such as heating, air conditioning, 12V battery charging, power brakes and power steering, as well as additional drive power and extended range.

The driving experience is transparent to the operator. The truck is started and driven in the normal way. The primary change in the driver interface is an additional electric accelerator pedal (pot box) mounted in line with the original accelerator pedal as shown in Figure 1.1.

In operation the driver's foot contacts the electric accelerator first, and the electric propulsion system is called on to deliver whatever torque it can. If more is needed for acceleration or hill climbing, the driver simply presses harder, and the gasoline engine, which is running all the time, adds the required torque. In steady driving and on gentle hills the electric drive provides all the propulsion power needed. The system works best with an automatic transmission with overdrive. The IC engine thinks the truck is rolling downhill all the way, with fuel economy to match.

We have achieved our technical goal of a drivable, capable supplemental PHEV that can deliver a 50% fuel saving or more with minimum cost of conversion. It has been developed on Ford F-150s, primarily 2WD Generation 11 (2004 to 2008) vehicles. We believe it will also fit Gen. 10 and the Gen. 12 and 13 models. It can likely be adapted to General Motors Silverado/Sierra trucks, Dodge Ram pickups and the corresponding full-size Nissan and Toyota models.

Project Evolution

This project began in 2006-2007 with conversion of a 1986 Chevrolet S-10-based van which had already been converted into a dual-fuel CNG-gasoline vehicle. This vehicle had a two-part drive shaft providing a convenient point of attachment for the electric supplementary drive at the center bearing. An Advanced DC Motors FB1-4001 9" motor was added driving a sprocket on the front universal joint yoke of the rear drive shaft with a 2:1 speed reduction[ii]. The sprocket was mounted on an overrunning clutch such that the electric motor could drive the vehicle, but would not be windmilled when the vehicle was being driven by the IC engine. A timing belt quickly proved inadequate, and an ANSI No. 50 chain was substituted. The center bearing was reinforced to resist the side load of the chain drive. Electric power was supplied by a bank of ten Group 24 AGM batteries and a Curtis 1231 controller. Lead-acid batteries were used for development to cut costs, with the intention to switch to lithium-ion when successful

This system functioned well enough to compete in the 2007 21st Century Automotive Competition "Run-to-the-River". This involved a 24 mile tour clockwise in the morning followed by a counter-clockwise run in the afternoon simulating a commute to and from work with charging at work. The S-10 had a top speed on electric power only of around 35 MPH on the level and an electric range of around 20 miles. The chain drive was effective but very noisy. Incidentally the CNG conversion was very satisfactory. CNG-electric could be a very desirable hybrid combination with a home-based natural gas compressor to utilize the existing infrastructure.

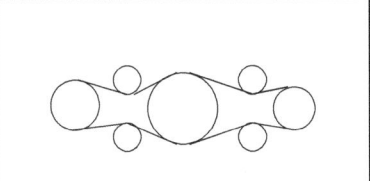
For 2008, a new drive with two 7" Advanced DC X91-4003 motors was constructed. The motors were suspended from cross bars so they could swing freely to tension a single 2" timing belt that looped over both motor sprockets and through a set of four idlers as shown in the sketch to provide adequate wrap around the driven sprocket[iii]. The belt was tensioned by a rod pushing the two motors apart, but allowing them to find their own positions, which were then held by a turnbuckle.

This drive was quieter and more satisfactory than the first one. It came in second in the 2008 Run-to-the-River to Jerry Asher in his early plug-in Prius. The fuel consumption was a somewhat disappointing at 24 mpg vs. the original unmodified value of 18 mpg.

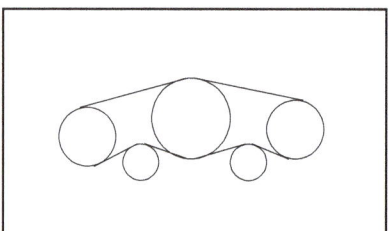

For 2009, a 1997 manual transmission Ford F-150 was converted using the same two motors in a slightly different configuration dictated by the limited headroom above the drive shaft as shown. The single overrunning clutch was expensive and ultimately failed due to over speed, so two smaller clutches were employed, one on each motor shaft. This meant that the drive was engaged at all times, but the motors did not windmill when not powered. A Gates Poly Chain was used in place of a timing belt for greater torque capability. The idlers were smooth faced and bore on the reverse side of the Poly Chain, which is acceptable, but not the most desirable arrangement. By the end of a year one chain had been used up in various mishaps and the replacement was showing wear.

Figure 1.2. Cross bed tool box with power electronics and charging cord. Batteries underneath.

The controller, charger, contactors, fuses and wiring were contained in a cross-bed toolbox with the batteries below, as shown in Figure. 1.2. Power and cooling air were fed to the motors from the box through flexible conduits leading through the sides of the bed to junction boxes shielding the brush ends of the motors. Control and instrument wiring followed the right hand power leads to the junction box on the right hand motor and on to the accelerator pedal. An ammeter and a voltmeter were mounted on the dash to provide information on power and energy usage.

This conversion incorporated a Programmable Logic Controller (PLC)[iv] to disable the electric drive when the clutch was depressed to allow for starting the IC engine. It also disabled the drive when the transmission was in reverse and when the brakes were applied. A voltage signal from the propulsion battery was monitored by the PLC to shut down the system when the battery was nearing exhaustion.

This conversion was more successful. It took first place at the 2009 21st Century Automotive Challenge in the Independent Light Duty Local category with a fuel efficiency of 28.1 mpgge, and an actual measured fuel consumption of slightly over 30 mpg. It was used by its owner for a year in regular driving.

In regular service the mileage was substantially poorer, and the truck was not a pleasure to drive. The manual shift, together with the clutch interlock and the safety system in the controller that prevents starting under load, required double pumping the accelerator pedal to get the electric system to restart after each gear shift, and the performance under electric power was nothing to write home about. The drive suffered from vibration and noise, and a number of minor fixes were tried without notable success. All in all, the owner was relieved to return the electrical equipment and get his truck back in original condition, in accordance with our agreement.

Figure 1.3. Single motor drive from the motor side.

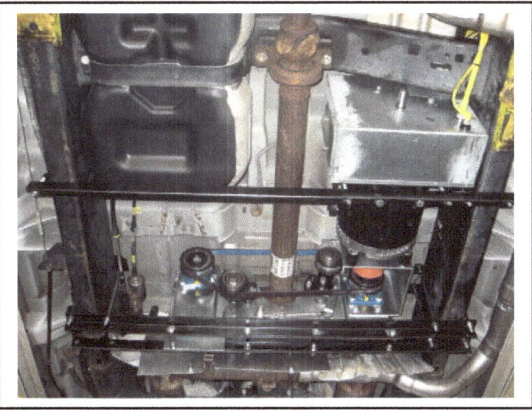

Figure 1.4. Drive installed in the vehicle.

Current Vehicle

To provide a more satisfactory conversion the 2004 Ford F-150, which is pictured on the front cover, was purchased. It has a 4.6 liter V8 engine and an automatic transmission with electric overdrive. A single-motor conversion [v]was developed, partly because many trucks only have room for one motor underneath, and partly to be able to use single-motor AC drives such as those from AC Propulsion and the Warp11from Netgain.

Figure 1.5. Driving Poly Chains and main sprockets on drive shaft. Heat shield removed.

The new drive used a single FB1-4001 Advanced DC 9" motor and a cross chain on the motor side shown in Figure 1.3 splitting the power between two driving sprockets to maintain balanced forces on the drive shaft The drive is shown mounted in the truck in Figure 1.4. Idler sprockets maintain the tension on the cross chain and allow adjustment to synchronize the cross chain with the driving chains on the forward side of the drive. Two driving chains are used to get better wrap on the driven sprockets. The balanced Poly Chain arrangement with two 12mm chains matching a single 21 mm chain is shown in Fig. 1.5 from below.

This arrangement is much more compliant laterally than previous drives and allows the drive shaft to move freely in both vertical and lateral directions to accommodate vibration of the IC engine and transmission on their flexible mounts. This has eliminated most of the vibration and noise encountered previously. The innovation has achieved the objective of imparting a pure electric torque to the drive shaft while balancing all the other forces and permitting the driven member freedom to vibrate freely. It also relieves the necessity for precisely aligning the supplemental drive with the existing drive shaft, because the drive can accept some misalignment as well as vibration.

Single Motor Performance

The single motor conversion was completed the day the 2010 21ˢᵗ Century Automotive Challenge began, and as a result we did not do too well. The number of entries in our class increased to four, and as it happened, none of them finished below us. On the other hand, it was a considerable achievement to be there at all, and we did achieve some improvements. Our fuel efficiency increased to 39.72 mpgge and the estimated petroleum displaced increased to 103 gge per year from 77.5 for the 1997 truck.

The driving experience with the single motor and the automatic transmission is perfectly acceptable. The truck is started and driven just as always. The electric assist is transparent and trouble-free. The system extracts the energy available in the battery and then shuts down automatically, but the trip continues without a hitch. A number of multi-run tests were recorded in 2010 and early 2011 during which continuous fine tuning took place. The results are summarized in Table 1.I below. Each line is a summary of the results from the previous refueling to the refueling that occurred on that date.

Table 1.1 Fuel and Electricity Consumption -- July 2010 through February 2011

Date	Miles	kWh	Gallons	Mpg	kWh/mi.	Notes
7/5/10	470.5	214.4	20.07	23.44	0.46	Mostly on I-95
7/19/10	437.7	212.3	18.30	23.92	0.49	Extreme range 34 mi. @40-45 mph
8/22/10	490	241.76	21.708	22.57	0.49	Improved drive with lightweight idlers
9/9/10	523.5	245.17	23.064	22.70	0.47	
11/9/10	180.4	151.63	4.779	37.75	0.84	Uprate to 200 V and increased ratio to 2.22 from 1.77.
12/22/10	187.7	179.72	5.736	32.72	0.96	
2/17/11	250.1	218.68	7.646	32.71	0.87	

The fuel consumption of the original truck was measured as 17.15 mpg, the average of two refuelings at 16.3 and 18.0. After addition of the batteries and the electric drive, the consumption without electric assist was 17.4 mpg, very little different.

In the July, 2010, runs, over-temperature and cut back problems were encountered with the Curtis 1231 controller, particularly at high speed on I-95. Installation of a massive heat sink, better cooling fans and current limiting to 220 Amps helped but did not entirely solve the problem with ambient temperatures in the 90s. Performance was somewhat marginal requiring constant blips of the IC engine throttle to maintain even 50-55 mph. The fuel mileage was a disappointing 23.44 mpg.

Running on secondary roads at 40-45 mph improved the driving experience in that the electric motor could maintain speed for long periods, but the fuel consumption was still only 23 mpg. Electric range was an encouraging 30-34 miles.

In late July an improved drive with lightweight idlers was installed to improve dynamic response and reduce vibration. Performance was still marginal on electric power, and the fuel consumption did not improve. The original battery pack consisted of twelve Exide group 27 sealed AGM batteries. These began to show serious loss of capacity during the summer driving season, and after extensive discussion with the manufacturer it was decided that AGM batteries simply cannot take repeated discharge cycles at the 2C rate. The pack was replaced with twelve Trojan 1275 flooded batteries with a remote watering system allowing the water in all 72 cells to be topped up from three points.

In October and November the battery pack voltage was increased from 144V to 196V by addition of four Valence U24-12XP lithium-ion batteries in series. These batteries were warranty rejects with failed temperature sensors, but they still had enough performance for a test of higher voltage. A separate Zivan NG-1 charger was added to charge the lithium batteries separately from the main pack, but still metered by the same 240 V, single-phase supply. The Curtis 1231 controller was modified to accept 200 V with uprated capacitors and MOSFETs By Dave Mosher of Cedar Rapids, IA.

At the same time another significant improvement in the drive was made and the sprocket ratio was increased from 1.77:1 to 2.22:1 to take advantage of the increased voltage. The rear axle ratio of the F-150 is 3.5:1 and the FB1 motor was now running at 6211 RPM at 65 mph. We thought we knew from bitter experience that the FB1 would blow up at 11000 RPM but hold together at 9000, so we still had some margin.

The performance with these modifications was spectacularly better than before. The truck was driveable on electric power up to 65 mph on the level and needed IC engine boost only for acceleration and steep hills at high speeds. The fuel consumption was finally at our target of a 50% reduction, albeit at the expense of a fairly heavy consumption of electricity. However, with 5.4 kW of solar power I was saving money and significant petroleum.

Unfortunately, currently manufactured FB1s won't take 6200 RPM, and we blew another one. It was replaced by a Netgain Warp 9" motor specially wrapped with Kevlar by Jimerico of Redmond, OR, to keep the commutator together. We limited the speed to 5500 RPM (60 mph) to be on the safe side.

While redoing the motor, we redid the entire drive with all the minor improvements which had accumulated over the eight months of running shown above, making a much more sound and producible set of parts. We also added a 6/500 electric clutch from Mayr Corp. to allow the truck to run on gasoline without windmilling the motor.

All of these modifications took us until the day before the 2011 21st Century Challenge, which was a fiasco. As in 2010, it was an accomplishment just to be there and compete, but the clutch installation (not the clutch itself) failed, and our mileage was horrible. We have since gone back to a clutchless installation, and the performance is back where it was. It might be possible to make the clutch work, but we have concluded that windmilling is more tolerable than the fixes needed to avoid it, and the penalty in fuel consumption is negligible.

Comparison of 144 and 200 V Operation

In August and September 2011 a definitive set of runs was made comparing operation at 200 V and 144 V. The truck was run over two separate routes, one mostly on I-95 and the other on local routes, mostly at 35-45 mph.

The I-95 route comprised 1.5 miles of 25 mph in town followed by 9.9 miles of Interstate including a 200 ft hill and 1.5 miles of 45 mph local roads. The return increased the end segments to 2.0 miles each because of the length of the freeway exits. The total round trip was 26.7 miles with 19.7 miles of Interstate driving.

The local route began with 1.0 miles of 25 mph local streets followed by 8.8 miles of 35-45 mph county roads, including the down-hill half the 200 ft. hill. This was followed by 2.9 miles of I-95 including the uphill half of the hill and 1.5 miles of 45 mph county roads, as before, total 14.2 miles. The return added 0.4 miles due to the extra freeway entrance for a round trip total of 28.8 miles. There are five stop signs and twelve traffic lights on each leg of this route, making it typical of suburban driving.

Table 1.2 Fuel and Electricity Consumption – August and September 2011

Date	Miles	kWh	Gallons	Mpg	kWh/mi.	Notes
8/13	206.8	148.6	8.33	24.83	0.62	200V Local route
8/22	228.7	123.34	9.42	24.28	0.54	200V Interstate route
9/8	506.9	254.17	22.06	22.98	0.54	144V Local route
9/16	220.0	123.0	9.27	23.74	0.56	144V Interstate route
9/23	395.3	90.39	19.69	20.1	0.23	144V Interstate route plus long trips.

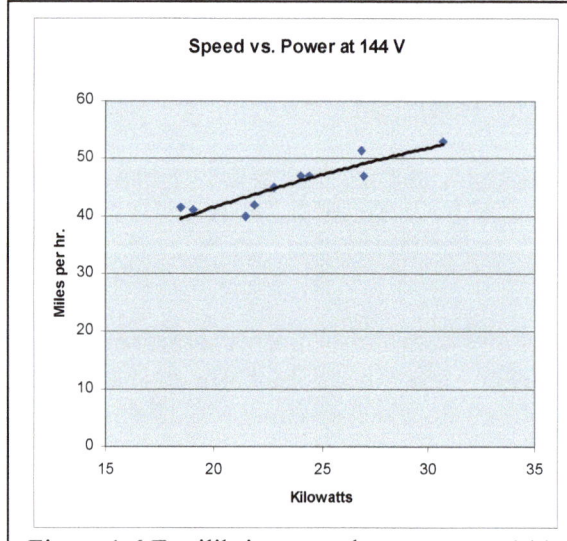

Figure 1.6 Equilibrium speed vs. power at 144 Volts.

The results at 200 V are not as good as those in Table 1.1 from the previous year. This is probably because the off-spec lithium batteries were on their last legs and were not able to complete the route on electric power, as shown by the lower amounts of electric power per mile used for the trip as a whole. The last 5 miles typically were on gasoline only, which leads to an estimate that the hybrid operation was at about 28 mpg with little difference between operation on the highway and on local roads.

The 144 V operations were even worse. This is not unexpected since there is less electrical power to supplement the gasoline. The equilibrium vehicle speed at 144 V and various power levels is shown in Figure 1.6.

The maximum speed that can be maintained on the level at 144 V is 52 mph. vs. 65 mph at 200 V. This is not an absolute bar to hybrid operation on the Interstate because the truck will maintain higher speeds for considerable periods of time, especially on a down grade. The technique is familiar as "hypermiling". The vehicle is accelerated on gasoline and electric power to say 65 mph and then slowly coasts down to some lower speed under electric power only, whereupon the acceleration is repeated. Clearly the higher the voltage and the more electric power available, the longer the coast, the better the gasoline mileage and the better the driving experience will be. Again, lead-acid batteries are only marginally satisfactory.

The September 16-23 runs were on the Interstate route with a long trip to Kempton, PA thrown in. This is representative of mixed use between commuting and out of town trips with the same vehicle. There are still savings in gasoline consumption when the electric drive is engaged, but the overall mileage is lower.

Lithium-ion Battery Results

Because of the disappointing results with lead-acid batteries a new lithium-ion pack was installed in late 2011. It consisted of 64 CALB 100 Ah cells in eight groups of eight with a nominal voltage of 205 V for a total capacity of 20.5 kWh. This is about the minimum that is practical for the F-150. Higher capacity cells can be fitted into the same configuration to get as high as 180 Ah and 36.9kWh. The groups of cells were connected with "Reg Decks" from Manzanita Micro, which are large printed circuit boards carrying the buss bars needed to connect eight cells in series along with all the voltage measuring connections for the BMS regulators to each cell and a thermistor to track the temperature of each cell. The regulators for each group of eight cells plug into sockets on the Reg. Decks. The use of these parts simplifies construction of the full pack enormously and makes a neat appearance. The cell voltages and temperatures are monitored by a display unit in the instrument module.

At the same time the entire electrical enclosure was rebuilt because experience with boats showed that the Reg. Decks and regulators were very vulnerable to liquid water. Since they are permanently connected to the cells, a single drop of water across a positively and negatively charged pair of conductors means instant and fatal corrosion. The solution was to enclose both the batteries and the power electronics in a bigger aluminum box rather than having the batteries underneath a tool box containing the power electronics. This arrangement has worked very well. The results of runs in late 2011 and early 2012 are shown in Table 1.3.

Table 1.3 Fuel and Electricity Consumption – Dec, 2011—March. 2012

Date	Miles	kWh	Gallons	Mpg	kWh/mi.	Notes
1/10/12	307.8	177.3	11.82	26.03	0.505	205V Local route
1/24/12	456.6	262	15.53	29.4	0.62	205V Yardley local route
2/26/12	132	110.8	3.53	37.4	0.84	205V I-95 route. Compliant drive, motor overheating
3/7/12	93.9	56.4	3.33	28.2	0.59	205 V Yardley route Compliant drive, Motor not overheating. Drive failed.

The performance with lithium-ion batteries is enormously better than with lead-acid. The maximum range measured was 37 miles with some in reserve in the dead of winter, whereas it had dropped to less than 20 miles with lead-acid. The truck has far more power and can climb hills equivalent to an intermediate ski run on electric power only. The batteries are rated at 400 A continuous, but are being limited to 300 A to lessen the stress on the motor and the drive. This is sufficient to reach 65 MPH on the level, equivalent to a somewhat excessive motor RPM of 6000. Almost all regular driving can be accomplished on electric power only. Acceleration is still marginal, and the gasoline engine is necessary to achieve satisfactory performance. However, with both gasoline and electric there is a noticeably faster take-off from a standing start than with gasoline alone. The truck feels livelier.

The first set of runs shown in Table 1.3 was on the local route with a couple of trips to Burlington, NJ, on gasoline only, and the mileage was not exceptional. The second set was on a new local route through Yardley adding a mile of 25 mph and eliminating the section of I-95 in the original local route. The total round trip distance is 31.5 miles with two extra traffic lights. The 200 ft. hill leading up from the Delaware River was climbed on electric power only, and the mileage was close to 30 mpg including 44 miles on gasoline only. The mileage on the electric portion was 31.8 mpg, a very satisfactory result.

Torsion Spring Drive

Figure 1.7 Compliant torsion spring added to cross chain sprocket to eliminate noise due to lack of synchronization with the direct drive shaft which had an equivalent spring on the driving chain side for balance.

The drive at this time was free of vibration but was noisy due to lack of synchronization between the cross chain and the driving chains. Torsion springs were added to the drives as shown in Figure 1.7. This eliminated the noise. A set of runs on the I-95 route gave excellent performance and exceptional gas mileage of 37.4 mpg. This is offset by the fact that the consumption of electric power increased from 0.62 to 0.84 kWh per mile, as one would expect. At high speed one is using more electric power to reduce the trip time during which the gasoline engine is idling.

The equilibrium speed reached with lithium-ion batteries is shown in Figure 1.8. It is very similar to Figure 1.6 in terms of speed per unit current, but this translates to 10 kW more power at the higher voltage. The reason for the increase in power at a given speed is not yet known. These measurements are difficult to make. As the speed increases the power increases at constant current, and vice versa, tending to make the speed unstable. The dispersion of the data is an indication of this. A corollary is that it is desirable to accelerate to slightly above your desired speed and let the electric drive maintain it at a higher level than it would if approaching from below. As expected, the curve extends to significantly higher power and higher speed with the added voltage.

Figure 1.8 Equilibrium speed vs. power with 205 V lithium-ion battery.

Motor Overloading

The added power and speed come with a penalty however in that the motor is seriously overloaded. The Advanced DC FB1-4000 which is very similar to the Netgain Warp 9 is rated at 30 HP continuous at 144V. At 205 V and 65 mph the Warp 9 motor was drawing 300 amps and putting out 73 HP. Not surprisingly it began to smell. This turned out later to be from overheated epoxy on the reinforcing windings on the commutator. The windings completely disintegrated, but fortunately the motor survived, though showing signs of

Figure 1.9. Failed torsion spring.

overheating. This is somewhat puzzling since the outside casing of the motor never got above 60 C and the overheat switch never opened, but clearly a 143% overload is not desirable or even sustainable. Massive brush wear and brush overheating were found on overhaul.

An additional set of runs were done on the Yardley low speed route to see if the motor could cope with this condition. The smell was absent and the motor seemed cool enough to the touch, but the compliant drive units failed in a relatively short time. The welds holding the springs onto the end collars failed and one of the springs distorted at the end as shown in Figure 1.9 A mechanical connection to the springs would be preferable to welding and stress relief. A differential gear assembly to apportion the single motor torque equally between the two driving sprockets as shown in **Drawing 23** would be still better in that it can accept any degree of differential motion between the two sprockets without adding differential torque. While a single 9" DC motor cannot drive a full size pickup truck, a single 11" Warp motor from Netgain should be able to, as well as the AC Propulsion drive, as shown in **Drawings 19 and 20**.

Light Twin Motor Drive

Figure 1.10. Twin seven-inch motor drive.

As an alternative to both the problems with motor overheating and failure of the torsion spring drive, a reworked twin motor drive was installed utilizing all of the experience gained since the original 1997 truck installation. The drive is shown in Figure 1.10. It used the original 7" motors, which are not really adequate, but were available for an inexpensive test vehicle. The motors were wired in series, which has the advantage that their torques should be perfectly matched under all circumstances since the same current is flowing through each motor. In this case the motors are slightly underloaded since the 205 nominal pack voltage results in a 103 V supply to each

motor relative to its 120 V rating. The triple driving sprocket arrangement was replaced by two 21 mm wide Poly Chains with a 71/34 tooth, 2.08:1 reduction ratio. With this drive there was no noise at any speed, though there was a so far unexplained vibration at 20 MPH. The performance with the light twin drive is summarized in Table 1.4.

Table 1.4 Fuel and Electricity Consumption--Light Twin Drive – April and May, 2012

Date	Miles	kWh	Gallons	Mpg	kWh/mi.	Notes
4/27/12	297.7	120.23	12,889	23.1	0.40	205 V Yardley route
5/17/12	235.1	94.37	9.045	25.99	0.39	I-95 route
5/18/12	56.7	21.15	2.90	19.55	0.37	21st CAC Range event
5/19/12	48.0	16.45	2.3	20.9	0.34	21st CAC Errands event

The maximum speed on the level with the light twin installation was 47 MPH at 26 kW (130 Amps) input, although, as mentioned above, it is difficult to measure this, and higher speeds can be maintained for considerable periods of time. It was enough to maintain 35 mph on the test track at Penn State for an electric-assisted range of over 50 miles.

The performance around town at 45 mph is adequate, but on the highway and on the difficult Errands Event route at Penn State, the light twin drive does not provide adequate power for acceptable driving or optimal fuel savings.

The Big Twin Drive

Figure 1.11 Big Twin Drive

Because the light twin drive was promising in terms of eliminating the vibration and noise of the single motor drive, and has inherent advantages of simplicity and torque balancing in those applications where there is room enough to deploy it, a larger twin drive was designed for two Netgain Warp 9 motors as shown in **Drawings 13-15** and Fig 1.11-1.13.

Figure 1.11 shows the drive complete on its 2x10 cradle allowing it to be balanced on the transmission jack for installation. The total weight is over 400 lb, which the jack can handle, but it should be treated with great respect. Balance the drive carefully with its center of gravity in line with the hydraulic cylinder of the jack, and with both the cradle and the drive cross member securely supported and strapped down to avoid accidents.

Figure 1.12 shows the truck with the larger 80 tooth x 21 mm double sprocket assembly mounted on the drive shaft and the exhaust rerouted to twin mufflers. This wasn't absolutely necessary, but is a much more convenient arrangement allowing better access to the drive, easier maintenance (and a more pleasing exhaust note.) Note the foil wrapped gasoline feed and return lines in the upper left. The Big Twin drive clears the gasoline filter mounted on the left frame rail with room for removal and replacement of the filter when necessary. It is necessary that the left

Figure 1.12 Big Twin 80 tooth sprockets and twin exhausts installed.

Figure 1.13 Big twin drive installed. Front view.

drive sprocket be the rear one to clear the fuel lines. For extra clearance a distance piece was added to move the sprockets back another 7/8 of an inch, but this brought the rear of the Warp 9 right up against the gasoline tank. **Drawing 9** shows a tighter arrangement of idler sprockets which allows ample clearance from the fuel lines and the rear of the motor. It will be desirable to remove the rear shaft of the motor and add a shield around the left sprocket to ensure that these components won't be driven into the fuel system in a collision.

Figure 1.14 Big Twin Installed. Rear view.

In Figure 1.13 the turn buckles which tension the driving belts may be seen along with the cables and levers which move the idlers outwards. Because headroom is limited at the midpoint of the motors by the floor behind the seat of the truck, the 9" motors are below the centerline of the drive shaft by 51 mm. This requires smooth idlers to bend the belts and locate the tensioning idler assembly concentric with the drive shaft. One of those smooth idlers can be seen on the right. It is critical that these not come loose even though they are rotating in a sense to unscrew their mounting bolts. To prevent that the bolt heads were drilled and a tie wire, which can be seen by its copper plating, is used to positively prevent rotation.

Figure 1.14 shows the installation from the rear. The formed cross member which supports the drive can be clearly seen. It is supported by a clip above the right frame rail, and a bracket which passes outside the left frame rail, clears the wiring mounted on the outside, then bends inward over the top of the rail. The rear of the motors is supported by a 2" x 1" (51mm x 25 mm) channel which just fits the space between the frame and the body. The rear ends of the motors are supported by straps as shown. This provides maximum ground clearance and a simple, clean installation. The series wiring from the controller to the right motor, then the left motor, then back to the controller can be seen.

The driveability with the larger motors is much improved, as expected. The top speed on the level is 62 MPH with the transmission in neutral. With the transmission in drive the top speed is 58 MPH, which is comparable to all the previous data. The F-150 does not freewheel in overdrive and there is a significant performance gain at high speed in shifting to neutral and letting the engine idle. This was done for all the data after 7/15/12. In neutral, the truck can be driven on the interstate at 55-65 mph with no input from the IC engine and can climb a 7% grade at 50 MPH on electric power only.

The performance of the Big Twin Drive is shown in Table 1.5. The motors are rated at 5500 rpm for one hour, which is equivalent to 65 mph with the 2:1 ratio between the 40 tooth drive sprockets and the 80 tooth drive shaft sprockets. This is an approximate match to the maximum speed they can attain except on steep hills. They are running at 150-160 Amps, which is just under their continuous rating of 180 Amps at full speed, which is also a good match, and cured the overheating and excessive brush wear observed when one motor was doing all the work.

Table 1.5 Fuel and Electricity Consumption-- Twin Warp 9 Dive
Summer, 2012 and Winter-Spring, 2013

Date	Miles	kWh	Gallons	Mpg	kWh/mi.	Notes
7/6/12	233.4	141.4	4.91	47.54	0.61	Hot weather testing, local
7/15/12	218.3	86.65	8.33	26.21	0.40	Local route in neutral
7/25/12	50.1	28.35	0.818	61.25	0.566	I-95 route in neutral.
8/4/12	172.5	83.38	5.48	31.48	0.483	I-95 route in neutral
9/7/12	630.3	335.96	15.838	39.80	0.526	Summer on I-95 route with 16 mi of local side trips
2/11/13	727.2	434.7	22.08	32.94	0.595	Winter on I-95 route with 32 mi. of local side trips. Max range 40 miles.
3/1/13	144.1	65.62	5.407	26.65	0.485	Evnetics controller shakedown. Weather cold.
4/6/13	530.6	307.72	14.744	35.99	0.5799	Still cold. Last 7 miles on gasoline
4/17/13	275.8	144.79	8.530	32.33	0.525	Still cold, Last 8 miles on gasoline.

The last week in June and the first week in July, 2012, were in the 90's and the motor temperature stayed well below 60°C. A test of the battery temperature rise showed an increase from 38°C to 41° after twenty miles of driving and to 44°C after 33 miles. The CALB cells are rated to 55°C discharging and 45°C charging, so they are safe to run uncooled, even with high ambient temperature.

The fuel consumption was much improved relative to previous drives, although the data are somewhat mystifying. The big twin drive posted by far the best MPG figure yet obtained at 47.54 mpg in the first set of runs on the local route. The fuel consumption was halved at the same electric consumption relative to the single motor drive. Running the same route with the transmission in neutral resulted in a 35% drop in electric consumption but an 81% increase in fuel consumption, which is counter intuitive to say the least, although the latter numbers are consistent with later data, leading one to think that the July 6 numbers are a fluke of some sort.

A short 50 mile run on I-95 gave exceptional fuel economy of 61.25 mpg , but a longer set of runs gave a good but not exceptional 31.48 mpg. A final, definitive set of runs in neutral on the I-95 route with a few side errands on local roads gave 39.80 mpg over 630 miles. The range, if we had gone to the full 26 gallon capacity of the fuel tank, would have been just over 1000 miles. The electric consumption was 0.526 kWh per mile or 1.9 miles per kWh. The equivalence between electric and gasoline power, allowing for the inefficiencies of each, is approximately 15 kWh per gallon, so 0.526 kW equates to 0.035 gallons per mile. The gasoline consumption was 0.025 gallons per mile. At the original mileage of 17.4 mpg, highway, the consumption would have been 0.057 gallons per mile. The electric drive saved 0.032 gallons per mile at the expense of 0.035 gallons per mile of equivalent electric energy. This suggests that the supplemental hybrid electric drive is pretty efficient.

A series of runs was made to determine the speed vs. electric power consumption, the course was a two mile water-level stretch along the Delaware River coming in the middle of the usual 27 mile round trip to Yardley, PA, so the batteries were approximately half discharged. The results are shown in Figure 1.15.

The two trend lines for Southbound (downstream) and Northbound (upstream) are fairly consistent and may reflect a slight grade in the course. The results show a startling improvement

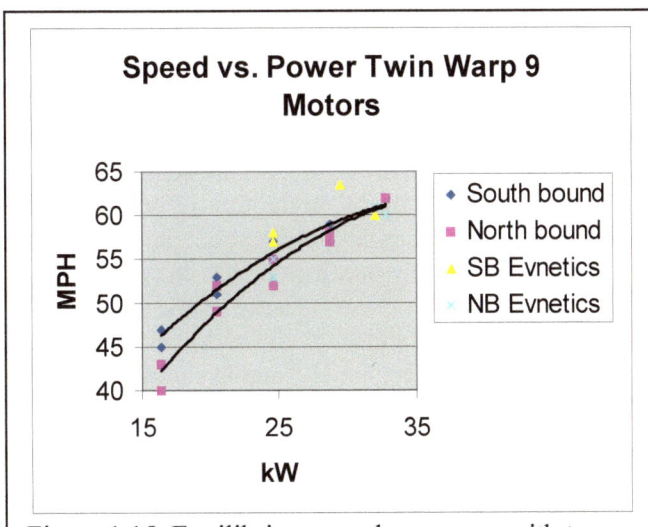

Figure 1.15. Equilibrium speed vs. power with two Warp 9 motors in series at 205 V.

in electric efficiency relative to the single motor drive results shown above in Figures 1.6 and 1.8. Previously at the rated 144 V a single 9" motor gave a speed of 52 MPH at 32 kW input, (220 Amps). At 205 V the single motor gave a top speed of 62 MPH at 60 kW, (300 A), a substantial overload. With twin 9" motors in series we are now getting 62 MPH at only 33kW, and 160 A.

A series of cold weather runs was performed in early 2013 to match the hot weather runs of 2012. The drive was modified slightly to overcome two residual problems with the otherwise perfectly satisfactory twin drive. One of these was a tendency for the Poly Chains to skip under high loads at low speed. Up to 30 MPH the counter EMF from the motors is so low that very large currents and torques can be generated which can overpower the chain drives despite the conversion to 40 tooth drive sprockets and increased tension. Because of the details of the idler supports, the sprockets were overhung by almost 75 mm (three inches) from the motors, and it was felt that this was allowing the shafts to flex and the chains to skip. In a redesign, the spacing was tightened up, as shown in **Drawings 13-15**, and the overhang was shortened to 38 mm. This did indeed improve the situation so that up to 200amps of motor current did not cause skipping, although higher currents at low speed still can. Since it is necessary to accelerate at low speeds with IC power anyway, a limitation to 200 Amps, (a 2C drain on the batteries relative to their limit of 4C), is not an inconvenience in the context of supplemental PHEV operation.

The other problem was a vibration in the drive between 40 and 50 MPH. This had been a constant with all of the conversions done to date, sometimes so severe as to be a show stopper, now reduced to a nuisance, but still present. It was associated with the chain drive. It was not present with the driven sprocket only and no chains. It was present with or without electric power. It was present when the IC engine was in neutral or in drive. It was greater when under electric power. It decreased as the sophistication and performance of the compliant drive system increased. The solution, better alignment of the driven sprocket, is described below under the Agni motor installation.

Yet another potential problem is wear on the reverse side of the Poly Chains running over the smooth faced idlers in the twin motor arrangement. Although all of the design specifications for idler size and power transmission are satisfied, some wear on the reverse side of the chains was seen in 3500 miles of operation. This also was eliminated in a twin Agni motor installation discussed below. The front face of the chains is in flawless condition with the blue paint of a new chain still showing and no evidence of uneven wear.

The winter runs in 2013 summarized in table 1.5 showed the boundaries of cold weather operation of the conversion. Fortuitously, the weather was unusually cold for New Jersey with temperatures well below freezing for most of the first two weeks. We found that the power from the battery decreased noticeably at a battery temperature of 0°C (32°F) and at -7°C (18°F) the battery voltage sagged drastically under power, representing the limit of useful operation.

Fortunately, the problem is self-correcting. The reason the performance is so poor is that the conductivity of the organic electrolyte in the lithium-ion batteries drops at low temperature. Because of the increase in electrical resistance and in I^2R heating in cold weather, as the batteries are used they warm up, and the performance is regained as you drive, at least above the low temperature limit. Typically the batteries would warm up by 5 or 7°C in the first 12 miles and an additional 5°C on the return trip reaching 10°C or more and essentially normal operation by the end of the trip. This effect also means that the range is relatively unaffected by cold weather, unlike lead-acid batteries and some lithium-ion batteries which loose capacity at low temperature. The maximum range in interstate driving in winter was indistinguishable from the summer value. This warming effect is more pronounced at low temperature, fortunately, and in summer at high temperature the electrolyte is so conductive that warming is limited to about 5°C, as mentioned above.

Charging is supposed to be limited to above 0°C, and in the winter it is convenient to charge immediately after the trip while the batteries are warm. Conversely in the summer it pays to delay battery charging till after sundown when they have cooled (and the load on the grid is reduced). If the batteries cannot be charged immediately or if the battery temperature is too low to run, a 150 W light bulb or other heater in the battery box for a couple of hours will bring them up to temperature.

A new Evnetics Soliton Jr. controller rated at up to 340 V and up to 600 Amps peak, 450 A continuous, was installed after the 2013 Winter series. A current limiting controller makes driving easier and is a more kosher solution for high battery voltage than the specially modified Curtis. In addition it has features that eliminate the need for the PLC. The inputs for brakes, reverse and park/neutral feed directly to the controller as well as a tachometer signal, if fitted, as shown in **Drawing 7.** Furthermore, the controller puts out 0-5 V signals proportional to battery voltage motor current and battery current providing safer instrument readings than by bringing the full pack voltage into the cab on the instrument wiring. Low battery voltage cut off can be programmed into the controller rather than requiring a separate signal and the optoisolator of **Drawing 7A.**

The Evnetics controller provided a major advance in driveability in that the motors never overpower the chain drive and cause it to slip. One can apply full electric power at any speed and supplement it with IC throttle as needed. During the first set of runs with the Evnetics ending on March 1, 2013, it took a while to get the controller programmed properly to deliver maximum performance and the gas mileage was not very good. The weather continued cold which did not help. Following the shakedown two extended runs were made successfully in weather that continued cold for New Jersey with night time temperatures in the thirties and daytime temperatures rarely above 50 F. The runs ending April 17 gave 35.99 mpg with an electric consumption of 0.58 kWh per mile, while the final runs yielded 32 mpg and 0.525 kWh per mile. These trips included some side trips and ended with some mileage on gasoline only which depressed the gasoline mileage. There was a short 27 mile trip on gasoline only which consumed 2.12 gallons for a mileage of 12.7 MPG indicative of the in-town cold weather performance of the truck itself.

Agni Big Twin Drive

Figure 1.16 Twin Agni 155R drive from the front showing idlers and 53-tooth driving sprockets.

The final (and recommended) drive configuration uses a pair of Agni 27 cm (10.62" OD) axial flux DC motors to provide a simpler, more compact DC twin drive with the same power capability as the twin Warp 9s but 250 lb lighter. This arrangement eliminates the reverse idlers as shown in **Drawing 8** and shortens the motor assembly, eliminating the need for a rear support as shown in **Drawings 9 and 10**. The entire drive fits under the hump in the floor under the seat. The unit and its installation are shown in Figures 1.16-1.19. Despite its large diameter, the Agni centerline can be on the same level as the drive shaft, eliminating the need for reverse idlers. These were needed to accommodate the longer 9" motors which were forced lower by a cross brace under the seat.

The Agni motor is a twelve-pole axial-flux design scaled up from the familiar Lynch/Etek/Perm/Briggs& Stratton 8"diameter 8-pole configuration. It is designated the 155-R because there are 155 radial conductors. It is rated at 23 kW and 62 N-m torque continuous duty at 120 V, 208 Amps. The motor is capable of 165 N-m of torque at 500 Amps for ten or twenty seconds, and can be run at 250 A, 82 N-m for one or two minutes. The maximum recommended temperature is 80°C measured on the stator, and we have come close to that on an 85°F day.

Figure 1.17 Sixty-seven-tooth driven sprocket assembled on front U-joint yoke.

Drawings 11 and 12 show the individual components of the Agni drive which are much like those of the other twin drives with differences in detail. A twin 9" installation can be modified to accept the Agni motor and vice versa. Aside from the reverse idlers, the main difference is that the Agni is limited to only 3600 RPM. The speed ratio is therefore 1.26:1 (67/53) to give a top speed of 64.3 MPH. The use of much larger driving sprockets makes for greater wrap and greater torque capability in the drive. A lower motor rpm may reduce noise and vibration. Brush life may be greater at lower speed but less with the axial design as the brushes have much less area, compensated by the fact that there are more of them.

The Agni is a permanent magnet motor with high-power magnets in the frame providing an axial field through a disc-shaped armature. There are twelve very narrow brushes, six positive and six negative, providing a current path through the armature. The speed is nominally 3600 RPM at 120 V, or 30 RPM per volt, and since it is a permanent magnet motor, it will not run away and blow up if it looses the load, as series-wound motors will. It will however generate full voltage and very high currents when driven at full speed, as it is in the PHEV application. A contactor in the motor circuit to isolate it when not connected to a power source is essential.

Overspeeding these motors is strongly discouraged as the armature windings are somewhat fragile, and the outer casing is sheet metal offering no protection if they disintegrate.

To avoid having to build a complete new motor mount, the twin 9" motor drive base was modified by drilling a pattern of holes to fit the Agni bolt pattern. Four of the six mounting screws could be engaged, and the 4" center holes for the Warp 9 motors provide the necessary cooling air flow. The mounting for the idlers is the same for both types of motors, but the Agni fits into the truck better. The result is a compact, simple drive with the Poly Chains contacting sprockets only on the toothed side as shown in Figure 1.18.

Figure 1.18 Twin Agni drive installed showing idlers and 67 tooth driven sprocket.

Figure 1.19 Twin Agni drive from the rear showing insulating plate left and junction box right side.

As a further simplification a single 62 mm wide driven sprocket was used as shown in **Drawing 12** and Figures 1.17 and 1.18. A new aluminum distance piece with the same 7/8" (22mm) spacing was made. Initial tests of the new drive produced very severe vibration from 40 MPH upwards, and we realized that the driven sprocket was cocked on the U-joint. This hypothesis was confirmed on disassembly by wear patterns on the 67-tooth sprocket showing clearly that it was not running true. Either the flange on the front yoke of the U-joint is not perpendicular to the shaft axis or the distance piece was hanging up on the fairly substantial radius from the shaft to the flange. To cure this problem a ring of holes was tapped into the distance piece for knurled point set screws, and considerable effort was expended to get the distance piece exactly perpendicular to the shaft as well as concentric. The result was a perfectly smooth running drive and elimination of the 40 MPH vibration that had been a problem for years. For future installations it will be important to mount the front yoke in a lathe to ensure that the flange and the distance piece are perpendicular prior to assembly.

Figures 1.18 and 1.19 show the drive installed in the truck. The drive has been further simplified by using a single turnbuckle to tension the idlers with loops of 3/32" wire rope on each side passing over a single pulley. Ball bearing pulleys are now used because the plain bearing pulleys tended to seize in service. The idlers are unmodified from the twin Warp 9 installation. Lighter idlers, as shown in **Drawing 12,** would be desirable since the experience with the 40 MPH vibration defines the limit of dynamic compliance of the present idler assembly. Up to 40 MPH the idlers could accommodate the eccentric motion of the driven sprocket, which is a pretty remarkable confirmation of the compliant drive concept, considering how terrible the alignment was in the last iteration. However, at higher speeds the idlers couldn't keep up with the eccentricity of the drive sprocket. Lightening the 2.3 lb idlers by half should raise the critical speed to 80 MPH and provide double protection against misalignment.

Figure 1.19 shows the installation from the rear. The Agni motors just exactly fit into the hump under the front seat of the F-150 vertically and lengthwise. Since they are so short, there is no need for a rear support. The Agni motors weigh 20 kg (44 lb.) compared to the Warp 9s at 76 kg (168 lb). The axial motors reduce the weight of the drive by 248 lb! It still needs a transmission jack for installation, but doesn't feel nearly as menacing at 128 lb total.

The Agni motors are wired in series as before to accommodate their limited voltage and to exactly balance the torque of the two motors. They were shipped without plastic covers on the brush assemblies, and during installation a potential short was created in the wiring of the left hand motor. During the first drive the potential became actuality, and because a permanent magnet motor can generate enormous currents, the Poly Chain drive locked up and slipped, while the arc started a fire in the cable insulation. Fortunately the fire was not severe, and the mechanical drive components survived. The brush holder and four of the brushes were replaced and a 5 mm plastic laminate insulation plate which can be seen in Figure 1.19 was added to prevent a recurrence. The insulation on the right hand motor is provided by a plastic junction box which mates with the 2" conduit coming down from the electronic enclosure and provides protection for all of the electrical and control connections. More robust cable connector posts were fabricated for both motors from brass as shown in **Drawing 12.** The performance of the twin Agni drive is shown in Table 1.6.

Agni Big Twin Performance

Table 1.6 Fuel and Electricity Consumption-- Twin Agni 155-R Dive
Summer and Fall 2013, Winter 2014

Date	Miles	kWh	Gallons	Mpg	kWh/mi.	Notes
June 3	334	181	9.415	35.48	0.46	I-95 route
June 12	237	102	7.561	31.33	0.436	Yardley local route
July 22	827	272	24.319	34.01	0.329	Includes a 100 mi. round trip to Phila.
Aug. 20	809	362	22.787	35.5	0.4479	Includes a 100 mi. round trip to Phila.
Sept. 18	800	351	24.032	33.27	0.4398	Misc. trips
Jan. 1 2014	1427	343	38.6	36.95	0.2486	Uprated battery to 230 V. 801 miles on electric 626mi. trips to Phila.
Feb. 12, 2014	303		19.272	15.72		Check gas mileage of truck without electric boost

Because our battery pack was initially limited to 205 V, we were not getting the full power and speed of which the Agnis are capable at their rated 120 V each, 240 V total. The top speed on level ground at 205 V was 57 MPH at a current draw of 150 A or 30 kW.

The acceleration was still marginal, but the performance in local driving was slightly better than with the Twin Warp 9 drive. The motors are capable of two or three times the continuous duty torque for short periods, which provides better acceleration, limited by the Poly

Chain rating of 56.9 HP compared to the 27 HP put out by each of the motors at continuous power 208 A. The drive can therefore tolerate a maximum of about 400 Amps. As mentioned above, with a motor current limit of 200 A, we were getting close to the temperature limit on the motors on not particularly hot days. However, since most of the heating occurs when the motors are up to speed and the current is limited by the back EMF, a higher limit of 300 A has since been used, and it does provide much more spritely performance It would be desirable to provide continuous temperature monitoring to cut back power in case of overheating. The Agni motors do not have over temperature switches, which would be a desirable addition.

The maximum high voltage battery pack that can be accommodated by the battery box of **Drawings 3 and 5** is eight strings of ten 130 AH cells each, totaling 256 V and 33 kWh. A less expensive alternative is to add a ninth string of eight 100 Ah cells for a nominal 230 V pack. This was done as part of a National Science Foundation Phase I Small Business Independent Research project entitled "Beyond the Smart Grid: Vehicle-Solar-Grid Integration. The resulting top speed on the level was measured at 61 MPH, somewhat disappointing since the expected speed based on the increase in pack voltage should have been 64 MPH. As discussed below, the performance of the CALB batteries was deteriorating after two years and 300 charge cycles. The fuel consumption, however, was very satisfactory at almost 37 MPG over 1427 miles of varied driving including a number of 100 mile round trips to Philadelphia as part of the NSF project. The fuel consumption without electric supplement was checked in January and February, 2014, and it averaged 15.7 MPG in the same type of service.

The fuel consumption in 3000 miles of testing in summer and fall weather under normal useage with a mixture of short and long trips was very satisfactory, being above 30 MPG consistently, a true 50% reduction of 0.029 gallons per mile from IC engine only operation. The corresponding electric consumption was 0.422 kWh/Mi or 2.37 mile per kWh, equivalent to 0.028 gallons per mile at 15 kWh per gallon equivalent. Again the electric energy just about balances the reduction in gasoline consumption.

In February, 2014 the brushes and Poly Chains were inspected for wear. The brushes had lost an average of 3 mm (1/8") out of a total available length of 12 mm (½") in 6000 miles of running with the Agni motors. A further 3200 miles resulted in another 2 mm of loss The total brush life is thus expected to be 22,000 miles, if the wear is linear, Fortunately, replacement is easy (though not cheap), particularly if the whole brush assembly is replaced as a unit, with the phenolic brush holder shipped back to the factory for reassembly.

The testing has continued throughout 2014 but with much more long distance driving to build and test an Electric Vehicle Photovoltaic (EVPV) installation at Penn State University's GridSTAR facility at the Philadelphia Navy Yard. The objective is integration of electric vehicle battery storage and solar photovoltaic generation into the electric grid (VSG) to allow the expensive battery to earn revenue for ancillary services to the Regional Transmission Operator, thereby offsetting the cost of the conversion.[vi]

Only 30-40 miles of these 100 mile round trips could accomplished on electric power, so the fuel consumption is higher than on the Yardley runs, but still little more than half the unassisted fuel consumption. The results are characteristic of normal use with a mix of long and short trips.

In September, 2014 the brushes were replaced at about 12,000 miles, as the earlier measurements suggested that they should be near the end of life. Indeed the wear was starting to become unequal with some of the brushes undesirably short. Replacement at 10,000 miles (every other oil change for the F-150) would be prudent.

Table 1.7 Fuel and Electricity Consumption-- Twin Agni 155-R Dive
Summer and Fall 2014

Date	Miles	kWh	Gallons	Mpg	M/kWh.	Notes
May 17	679	260	24.11	28.18	2.61	2 trips to Philadelphia
June 5	683	193	25.74	26.55	3.54	3 trips to Philadelphia
June 15	590	83	23.86	24.74	7.14	4 trips to Philadelphia
July 9	684	216	24.00	28.51	3.16	2 trips to Philadelphia
July 21	680	184	24.27	28.02	3.72	5 trips to Philadelphia
Aug. 16	976	329	33.00	29.57	2.97	5 trips to Philadelphia
Sept. 7	637	183	21.95	29.00	3.48	3 trips to Philadelphia replace brushes
Sept.29	695	211	24.60	28.25	3.29	1 trip to Philadelphia
Oct.29	549	171	18.80	29.22	3.21	1 trip to Philadelphia
Nov.17	1121	401	39.10	28.66	2.79	Two refuelings combined

This is the sole undesirable feature of the Agni motors. A complete brush assembly is £195 (British pounds) or about $330 times two for two motors. A set of brushes only costs £144 or around $250 per motor. This works out to 5 cents a mile, which is pretty steep. We may be able to get the life longer or the price lower with further development, but that is the situation now.

The Poly Chains appear to be in perfect condition, tracking well and with no signs of wear or stretch. No estimate of their expected life or maintenance interval is possible, but an inspection at 5000 mile intervals would be advisable.

Within the limits of speed and acceleration imposed by the motors, the twin Agni drive has proven itself very attractive after some teething troubles. It has been trouble-free for 15,968 miles of driving in all kinds of weather and all kinds of trips over almost two years. The drive is silky smooth at all speeds. There is a slight whine from the Poly Chains at low speed but silence at higher speeds. The truck maintains speed on hills and is a pleasure to drive.

Performance Summary

An F-150 with the Big Twin conversion can achieve at least 61 mph on electric power only. Higher speeds are not really desirable because of excessive battery draw and the limitations of the electric motor which is turning at 89 rpm per mile per hour. At 62 mph, the motor speed is 5412 RPM for the Warp 9 conversion and 3378 RPM for the Agni, which is close to their RPM limits of 5500 and 3600 RPM respectively for continuous operation. On steep downhills 65 mph has been successfully accommodated, but higher speeds definitely risk destroying very expensive motors.

The fuel consumption of the truck on electric power is cut in half. The original mileage was 17.5 mpg and under electric power this increased to 31-39 mpg. In competition at low speeds, a maximum reading of 39.72 miles per gallon of gasoline equivalent was recorded. The extended summer run on I-95 gave 39.8 mpg.

Plug-in Hybrid Pickup Trucks

With lithium-ion batteries, an electric range of 35 miles has been achieved routinely on the interstate. The calculated range with the nominal (20 hr rating) 20 kW hr battery pack is 40 miles at 40-45 mph, but the actual range is greater due to the fact that the maximum current is not drawn all the time. An extreme electric-assisted range of 55 miles was recorded twice in the 2012 and 2013 21st Century Automotive Challenge competitions under ideal conditions at a steady 35 MPH. With lead-acid batteries the range has been as low as 12 miles in the winter time, another reason why lead-acid is not really viable.

The driving experience with an automatic transmission is virtually unchanged from the original truck. The vehicle is started and run in the usual way. While it will accelerate from a standing start under electric power only, the rate of acceleration is painfully slow, and some IC engine input is welcome. Once at speed, the truck can operate for long distances and climb considerable hills on electric power only (50 MPH up a 7% grade), requiring gasoline supplement only for acceleration and climbing sustained hills at high speed. Without free-wheeling it is necessary to shift into neutral for best speed and fuel economy on electric power and shift back into drive for steep hills.

When the battery is nearing exhaustion, a droop in battery voltage and a decrease in current available to maintain speed will be noticed. When the droop becomes excessive the PLC or the controller will shut down the electric system and the trip can be continued seamlessly on gasoline. If it is necessary to travel long distances, the electric system can be shut down and the trip completed on gasoline. It is still necessary to limit the speed to 62mph to avoid blowing up the motors.

The thermal load on the motors and controller is safely below the upper limit. At 62 mph the current draw of the Warp 9 motors was 160 Amps compared to their continuous duty rating of 180 Amps. This should solve the excessive brush wear noted with the single motor installation. The power input was 32 kW and the motors were putting out 43 HP. The battery should be chosen to deliver 180 amps continuously at 200 V to permit maximum speed on the Interstate. The electric power consumption is roughly 0.40 kWh per mile. The performance with Agni motors is very similar, except that the current draw tends to be a little less.

For a twin motor drive there are four alternative motors. The Advanced DC FB1-4000 and the Netgain Warp 9 are essentially equivalent series-wound DC traction motors with the same outline and the same power output. The Hi Performance Electric Vehicle Systems AC 50 AC motor is very similar with the same shaft and mounting details but two inches shorter and 26 pounds lighter. With a Curtis 1238-7601 controller it can operate on 96 V DC and produce 50 HP peak, 25 HP continuous. It is more expensive than the DC motors but provides regenerative braking, higher speed capability, a 17% weight saving and no brush maintenance.

The Agni 155R axial flux permanent magnet motor is 27 cm (10 5/8") diameter by 125 mm (5") long weighs 20 kg (44 lb) and puts out the same power as an ADC FB1-4001 or a Warp 9 which are 16" long and weigh 168 lb. The units we installed in mid 2013 were preproduction and not entirely finished. They have performed very well in over 15,000 miles of driving in wet and dry weather over a full year. As is not surprising, with their very narrow brushes, brush wear is a consideration.

The performance of the single motor drive was encouraging, but not really ready for prime time. The mismatch between the front and rear belt drives, both of which are rigidly in phase with the drive shaft has not been successfully accommodated by automatic adjusters. The torsion spring compliant drive might be made to work, but a better alternative is the differential compensator shown in **Drawing 23.**

The 8mm pitch Poly Chain drive has been satisfactory when properly aligned and not overloaded. Alternatives have not appeared to offer any advantage. ANSI roller chain does not have the speed and power capability needed. It is noisy and requires shielding and lubrication. Silent steel link and HV chain have the speed and power capability but require pumped lubrication and are quite heavy and expensive. 14 mm pitch Poly Chain has ample power capability and marginal speed capability but the smallest idler sprockets are 28 tooth, which are too large and heavy for the twin 9 inch drive. A 14 mm 20 mm wide 30/38 tooth combination could be used with the twin Agni drive. The 8 mm drive has twice the power capability needed, providing adequate margin, and should exhibit satisfactory life to match its satisfactory performance.

The performance of the major electric components has been satisfactory. The Manzanita charger has been trouble free and reliable. It backs off the rate of charge if it overheats The only problem has occurred when the charger lost contact with the battery due to a defective contact and a blown fuse. When they loose the load, Manzanita chargers go to overvoltage and self destruct. They have internal fuses and should not be externally fused.

The Evnetics controller has been satisfactory and a great improvement over the uprated Curtis. The only quibble is that the current and voltage meter outputs are not the most accurate and need to be separately calibrated to produce useful information. The controller failed to operate at one point indicating low 12 V input, which was not true, but the problem corrected itself.

The CALB lithium-ion, iron phosphate cells have been a major improvement over even the best lead-acid batteries, but they are starting to show degradation after two years and almost 300 discharge cycles. One cell failed after about a year of service and another about a year later. A number of others are developing high internal impedance. The cells do not appear to have lost significant capacity but their power capability is reduced because of the increased resistance. This may explain why we did not get the full benefit of going to 230 V that we expected. This experience is not unique. Other users of CALB, Thunder Sky and the like cells have had similar problems. It may be wise to use Valence batteries, even though they are more expensive. We have had a set of Valence batteries in marine service since 2007. In Chapter 3 and the associated drawings we are offering instructions on both Valence group 27 batteries and CALB cells, which require quite different installations.

At its present stage of development the twin motor supplemental PHEV conversion is operable and satisfactory in daily service on local roads and on the Interstate. The target fuel consumption has been achieved, and the drive is free of excessive noise and vibration. With series-wound motors it has functioned over nine thousand miles without indications of excessive wear on the active side of the Poly Chains, although there is some wear on the reverse side. This wear has been eliminated at the expense of shorter brush life in the Agni version over 16,000 miles. Additional development may provide improvements in performance and durability, but the system is now ready for beta testing to probe its durability and maintenance issues.

Chapter 2: The Conversion in Detail

Component Technologies, What You Can Expect, Facilities, Cost, Savings

Major Component Technologies

Our objective in this conversion is to provide as much as possible of the advantage of electric propulsion with absolute minimum modification of the vehicle and therefore minimum cost. To do this we leave the Internal Combustion (IC) drive train completely intact and unmodified. The chief modification to the vehicle is to reroute the exhaust to provide room for the electric motors. This is done downstream from the catalytic converters and the oxygen sensors to avoid "tampering" with the emission control system, which would require recertification of the vehicle. A hole pattern is drilled in the vibration damper boss on the front yoke of the front drive shaft to connect the electric drive. Two brackets under the seats need to be removed to clear the drive. These are the only modifications to the vehicle. Everything else in the conversion is an addition. The conversion can be completed without ever opening the hood. The added components are grouped in major subsystems which should be pre-assembled for easy and quick installation.

The major added components are:

The battery pack, which stores the electric energy

The control system, which allows the driver to add electric power to propel the vehicle in any amount needed that the battery can supply.

The charger to replenish the battery from the electric grid via the plug, (as in Plug-in Electric Hybrid Vehicle, PHEV).

The motor and drive system, which converts electric power from the controller into torque applied to the vehicle drive shaft.

The Programmable Logic Controller (PLC) which interlocks the electrical system to avoid conflicts between the electrical and IC drives and protects the battery pack, (and which may be included in some controllers).

The Instrument package, which allows the driver to monitor the performance of the electric system.

The Battery Pack

The heart of any electric vehicle is the electric storage battery that supplies the energy to drive the vehicle. It is the gas tank, and the limiting factor in the performance and cost of the vehicle. As mentioned in the introduction to Chapter 1, new developments have revolutionized battery technology and made possible a new generation of PHEVs.

Battery Technology

To realize the potential of PHEV conversion a better battery than the familiar lead-acid cell is really necessary. This is no surprise. Conversions of all kinds based on lead-acid batteries have been around for decades without making a real impact.

Rechargeable electric storage batteries date back to Edison's time, and before. They have many uses such as starting and lighting vehicles, but have never had the capacity for satisfactory main propulsion. Early electric automobiles and boats were rapidly replaced by those powered by internal combustion engines burning petroleum fuels. Petroleum provides incomparable energy storage for high power and long range. The specific energy of gasoline is 48 million Joules per kilogram, or 13,333 Watt-hours per kg in battery terms. For comparison, the specific energy of the familiar lead-acid and nickel-cadmium batteries is approximately 35 Wh/kg on a good day (half that on a cold day, less still at high loads or after a year or so of service). Advanced nickel-metal hydride batteries are quoted at 75 Wh/kg.[vii]

To be sure, only 1/3 to 1/4 of the energy in gasoline can be used because of thermodynamic and mechanical losses in the Internal Combustion (IC) engine. The efficiency of the utilization of electric energy can be 80% or higher, but even 3333 Wh/kg actually applied to the road or the water from gasoline is a lot higher than 35 Wh/kg from batteries. It is astonishing that electric cars and electric outboard motors for boats have done as well as they have with this kind of handicap in their basic energy supply. Both are in use by the millions, as golf carts and trolling motors.

Lithium–Ion Batteries

Lithium-ion technology has been around since the late 1980s. As with microelectronics, the technology has been driven by demanding and fast-moving consumer electronic requirements. The demands of laptop computers and cell phones for high-performance, portable power sources are particularly challenging.

Lithium is desirable as an anode (negative battery electrode) material because of its low atomic weight and high single cell potential of over 3.0Volts compared with lead-acid at 2.2V and nickel-cadmium and nickel-metal hydride at 1.2V. In addition to a big boost in specific energy, this means that fewer cells are needed to achieve a given voltage. (A cell consists of a single anode-cathode pair. A battery consists of an aggregation of cells to achieve a desired output, such as the familiar six-cell, 12-Volt starting, lighting, ignition battery). Conventional batteries cannot use high-energy anode metals such as lithium because the electromotive potential of the metal is higher than hydrogen, and it dissociates water-based electrolytes. Lithium-ion batteries use an organic electrolyte that is ionically conducting but proof against decomposition in the presence of lithium.

Since lithium metal is prone to burst into flames in the presence of water or oxygen, it is a safety hazard if the cells are damaged, or if they are overcharged and rupture. An early introduction of lithium batteries with metal anodes in the 1980's led to a series of fires, which gave lithium batteries a bad reputation.[viii] The lithium-ion battery provides a solution by using a graphite anode into which the lithium ions can intercalate (insert themselves between the layers of the graphite structure) to the extent of one lithium to six carbon atoms (LiC_6). The graphite acts as a sink for lithium and as a current carrier to convey electrons to and from a copper-foil current collector.

On the cathode side also, the lithium ions do not react but rather intercalate into a lithium-transition metal oxide or phosphate. The cell is constructed with the two electrodes separated by a porous polyethylene-polypropylene separator saturated with the organic electrolyte containing lithium perflourophosphate as the ionizing salt. The cell is charged by applying a voltage to draw lithium ions out of the mixed oxide through the electrolyte into the graphite. Approximately half the lithium can be transferred at full charge without materially affecting the structure of either the cathode or the anode material.[ix]

Lithium-ion cells are unique in that the crystal structure of the electrode materials is unaffected by charge and discharge. All that happens is that lithium ions shuttle from one side to the other. This minimally disruptive process has been described as "rocking". Since there is no structural change in the electrode materials, the lithium-ion cell can in principle undergo an infinite number of charge-discharge cycles without material change in its properties. It is, however, very sensitive to abuse. Aqueous cells tolerate overcharge by simply electrolyzing water. Lithium cells have no such safety valve, and overcharge causes at least permanent damage and at worst a fire. Similarly, overdischarge results in irreversible damage to these very expensive cells. An electronic battery management system that can detect the state of charge of each cell and prevent damage is mandatory.

The acceptance of lithium-ion batteries has been rapid. From a research curiosity in 1990, sales reached an estimated market value of $1.86 billion in 2000.[x] Since then increasing emphasis has been placed on large format batteries for vehicle propulsion, along with emphasis on safety in a crash-prone environment. The winning technology is based on phosphate or spinel anode materials which sacrifice some performance (3.2 V/cell vs. 3.7V for cobalt oxide anodes) for much less potential hazard. This technology was pioneered by Valence Technology and Lithium Technology, but has been widely developed by A123 and a number of other US and Asian companies.

Batteries have the advantage, relative to other advanced energy storage means, such as high-pressure hydrogen, that they provide energy in its purest form as electricity. The output is directly convertible into heat, light, propulsion, air conditioning, electronic calculation and the myriad other tasks required on a vehicle or in the home with very high efficiency. No further energy conversion is necessary. Batteries are the ultimate energy storage medium, and lithium-ion may be the ultimate battery.

The power available from lithium-ion cells is enormous. AC Propulsion of San Dimas, California have achieved an astonishing acceleration of 0-60 mph in 3.7 seconds with their T-Zero competition sports car powered by 6,800 laptop-sized lithium-ion cells.[xi] A similar pack is used in the Tesla Roadster and the new Tesla Model S. The specific energy of lithium-ion cells can be as high as 160 Wh/kg. Typically lithium-ion cells are about the same weight, size and cost as nickel-metal hydride cells, which are the next best option, but provide three times the voltage and three times the energy. Nickel-metal hydride cells have an aqueous electrolyte and are tolerant of overcharging, providing a safety and convenience feature in undemanding applications. Where performance and/or cost are critical, however, the electronics for ensuring safety with lithium-ion cells will be justified. All of the new generation battery electric vehicles and PHEVs use lithium-ion batteries.

Life

Cost of ownership is determined not only by initial cost but also by life. The cycling and calendar life of lithium-ion cells has received a great deal of attention recently because of the interest in hybrid vehicles and aerospace applications. Aerospace offers a promising field for lithium-ion batteries because of their very favorable specific energy and specific power, as well as cold weather capability and energy efficiency. The Mars Rovers launched in 2003 one of which is still performing on Mars are the first application of lithium-ion batteries for space missions.

Fellner and coworkers[xii] at the Air Force Research Laboratory did long term cycle testing and life prediction on lithium-ion batteries for both Geosynchronous (GEO) and Low Earth Orbit (LEO) applications. The target is a ten-year life for GEO (20 eclipse periods/cycles) and 30,000 cycles for LEO. The results show that the shelf life and the cycle life can be very long under realistic conditions of less than full discharge.

Belt and coworkers[xiii] at the Idaho National Engineering Laboratory tested large, 55 Wh cells under hybrid vehicle conditions over 11 months. All of the cells tested achieved 120,000 cycles with only a 7.5% loss in capacity and met all of the goals of the test, independent of depth of discharge.

CALB (Chinese Aerospace Lithium Battery Co) claim a life of 2000 cycles (six years of daily trips) at 80% depth of discharge and at a 0.3 C (current divided by battery capacity in Amp hours) charge- discharge rate. Torquedo claims 800 cycles at 100% DOD. RFE claims more than 1000 cycles at 80%. Thundersky (Winston Battery) claims 3000 cycles at 80% and more than 5000 at 70%, all based on manufacturer's literature. (But see the end of Chapter 1 for real world experience in a PHEV conversion).

The conclusion of these and many other studies seems to be that if the lithium-ion cells are not pushed to the limit of their capacity, except when they have to be, the cycle life can be very long indeed. The objective of matching the life of the battery to that of the vehicle seems well within reach by simply limiting the charging voltage. GM, Nissan, and the other major manufacturers offering PHEVs and BEVs based on lithium-ion batteries are warranting the batteries for up to ten years under very stringent conditions imposed by the California Air Resources Board, and they can do this by limiting their charging voltage and depth of discharge to the middle 60-80% of nominal capacity.

Cost

Lead-acid batteries are the lowest initial cost option for the battery pack. The problems with lead-acid are weight and life. Maintenance-free Absorbed Glass Mat (AGM) batteries and gel cell batteries are **not recommended** for this application because the continuous heavy current drain of propelling a heavy vehicle is more than they can handle.

Flooded lead acid batteries, similar to those used in golf carts, can be used. Golf carts usually use 6V or 8V batteries, which are very rugged and low in price due to their large volume of sales. Unfortunately they are not suitable for this application because the minimum acceptable pack voltage is 144 V requiring twenty-four 6 V batteries weighing a total of 1440 lb and costing roughly $4000.

A pack of twelve Trojan 1275 12 V batteries costing $2500 has been used successfully. The task of adding distilled water approximately once a month was simplified by the use of a manifolded watering system. They gave adequate performance in a full size pickup but weighed half a ton, and that is a sizeable fraction of the payload.

The other problem is life. Nominally you should expect to get two years of use of a flooded lead-acid pack, something like 400 charge cycles. In fact, even with high quality batteries, you will be lucky to get one year, and may get as few as 100 cycles, due to the stress of heavy current drain and deep discharge cycles. This can be alleviated by adding capacity to reduce the current drain and operating at only 50% discharge rather than 80% but now the weight is a ton, and the cost is doubled also to $5000. This is not a feasible solution for an every-day working vehicle.

The cost leaders in lithium-in cells are the Chinese including Thundersky and CALB. Common sizes are cells in the 40 to 200 amp hour range. Typical prices are in the neighborhood of $300-500 per kWhr of storage capacity, plus an additional 30% or more for battery management electronics and grouping into high capacity vehicle propulsion packs. The cost of the raw materials suggests that the United States Advanced Battery Consortium (USABC) "Goals for Advanced Batteries for Electric Vehicles" of $150/kWh can be approached and perhaps reached with additional development and quantity production.

Valence Technology batteries are roughly twice as expensive as the individual cells from China, but this is offset by the fact that they are packaged in 12 or 18 V modules with the bulk of the BMS function included in a sealed, weather-proof unit. The saving in pack assembly cost, and the advantage of a more rugged installation may offer enough added value to justify the extra cost. The extra life that we have experienced with Valence batteries in marine service (seven years) certainly does.

Lead-Acid and Lithium-Ion Packs

One of the advantages of the supplemental PHEV concept is that it can accommodate battery packs of different sizes and get the most out of any of them. The minimum pack voltage for satisfactory performance is 144 V. This is conveniently supplied by twelve twelve-volt batteries. Figure 2.1 shows an early installation of twelve Group 27 AGM batteries grouped into three nominally 48 V modules. This pack provided marginally satisfactory performance in early testing but failed after three months in regular service. The manufacturer agreed, after the fact, that AGM batteries cannot stand a continuous 2C current drain.

Figure 2.1. Early lead-acid battery pack. Twelve AGM group 27 batteries divided into three 48 V modules totaling 144 V.

This pack was replaced by a pack of twelve Trojan 1275 twelve Volt flooded batteries. These lasted for two years of intermittent testing or approximately 100 charge cycles. They had the power to drive the truck as shown in the data in Chapter 1, but suffered from drooping current capability (High internal impedance) especially when nearing complete discharge.

Fortunately, the development of lithium-ion batteries for vehicles has matured at just the right time to solve this

Figure 2.2 Lithium-ion battery pack. Sixty-four cells grouped into eight 25 V batteries totalling 205 V.

problem. Twenty-first Century lithium-ion phosphate chemistry provides a battery with three times the specific energy storage of lead in kW hrs per kg, without the thermal runaway hazard associated with earlier cobalt-based laptop batteries, and with almost no heat generation on charging or discharging, thus requiring no cooling system.

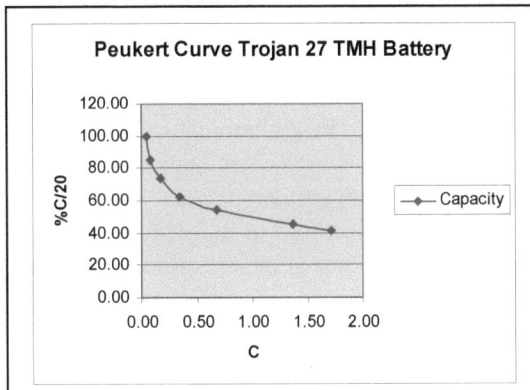

Figure 2.3 Peukert curve of amp hour capacity vs. current drain for lead-acid batteries from Reference ix.

Thus the recommended battery pack for the supplemental hybrid is 205 V or sixty-four 3.2 V cells in series with at least 100 amp hours of capacity as shown in Figure 2.2. The cost of this pack will be approximately $8000 and represents almost half the total cost of the conversion. Its life is not known with precision because of too little experience with this new technology, but even after their capacity has dropped to 80% or less of the original, the battery pack still has value as energy storage for electric utilities and some of the original cost should be recoverable.

Capacity

The battery drain increases from a small fraction of the motor current at low speeds to the limit set by the controller at mid speed say 300 Amps, to the limit of 180-200 Amps determined by the counter-EMF of the motor at high speeds. It is desirable to limit the battery pack drain to less than twice its Amp hour capacity (the 2C rate). Lead-acid batteries loose capacity rapidly with increasing current, (shown by the Peukert Curve in Figure 2.3)[xiv]. At 1C the capacity is about 50% of the C/20 rated capacity, and at 2C it is down to 40%. At 0°C (32°F the capacity of lead-acid batteries is reduced to 70% and at -18°C (0°F) it is 50% of rated capacity[xv]. Between the two effects the 35 kWh/kg can sink to 10 or less resulting in a range decrease to 30% of the nominal expectation.

Lithium-ion batteries are much less sensitive to current draw (and to low temperatures) suffering little loss in capacity up to 2C and down to 10°C. Some lithium-ion batteries are rated for continuous duty at higher currents. Valence batteries are rated for continuous duty at 1.36 C. Winston (Thundersky) says continuous service at 3C and CALB says 4 C. Thus for a 100 Amp-hour pack, the minimum practical size battery, the battery current should be held to an average of less than 200 A if possible, and never more than 400 A.

High performance cobalt-based lithium-ion batteries will heat up on both charge and discharge. Since they are susceptible to thermal runaway, they require cooling. Tesla for example uses liquid cooling to control their 6800 cell high power pack. This involves a considerable increase in pack complexity and weight. Fortunately, the safer iron phosphate cells show much less temperature rise and can be used uncooled in most applications.

The Charger

The battery charger to return energy to the Direct Current (DC) battery pack from the Alternating Current (AC) electric grid must meet a number of requirements to achieve optimum battery performance. This is particularly true for lithium-ion batteries, as described above. As a result lithium-ion batteries must have a charger which is under the control of a Battery Management System (BMS) which can prevent overcharge, and ideally over discharge and over current as well.

The charge cycle is almost universally divided into a constant current segment limited by the capacity of the charger, followed by a constant voltage segment limited by the maximum voltage tolerated by the battery. This may be followed by a float range during which cell equalization can take place and cell voltage is maintained indefinitely.

Modern battery chargers accomplish this level of control and much else besides. Virtually all multi-cell chargers are now of the switching power supply type. The AC input is rectified to DC in a full wave bridge. The DC is chopped (switched) to provide a high-frequency (25 kHz) AC current which passes through a transformer for rough voltage adjustment and an electronically controlled Pulse Width Modulated (PWM) current limiter to provide the desired output. This is then rectified and filtered to provide the final DC output at a controlled current and voltage. The control system can be programmed to provide any desired profile of constant current, and constant voltage vs. time. Sophisticated chargers can also accept battery temperature measurement and adjust the charging voltage accordingly.

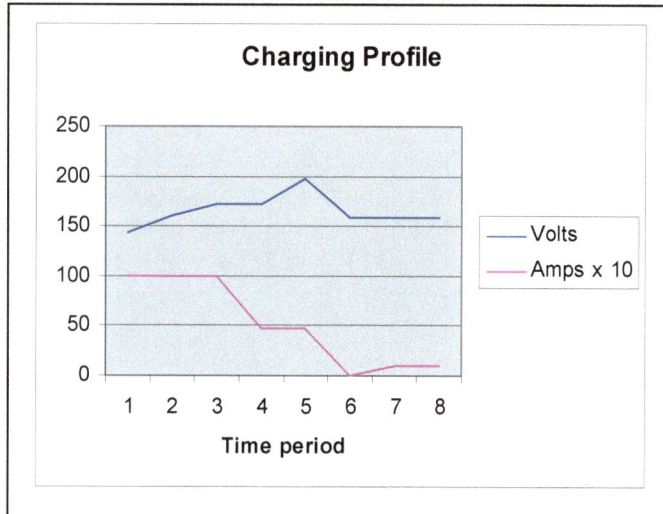

Figure 2.4. Typical charging profile for 144 V pack of lead-acid batteries provided by Electric Conversions and Zivan.

In the case shown in Figure 2.4.for a 144 V flooded lead-acid pack there is a constant current portion where the voltage rises to its set point of 171.6 V, followed by the constant voltage section in which the current drops to 4.8 amps. This current is continued for a specified time or until the voltage reaches its safety maximum of 198 V to equalize the batteries in the pack by bringing the least well charged of them up to full charge, while allowing the others to electrolyze water, producing hydrogen and oxygen gas. The equalization phase is followed by a drop to 159 V (13.25 V per battery), a float voltage which maintains the battery in a fully charged state for a specified time or indefinitely.

Keeping a lead-acid battery pack fully charged is important both to maintain its maximum capacity and because partially charged batteries deteriorate by sulfation of the positive electrode, converting lead sulfate into an unreactive crystal structure that cannot be reoxidized to lead oxide.

The same profile, less the equalization spike, can be used for lithium-ion batteries. Typically for nominal 3.2 V cells, the constant current section will charge the pack at the 0.3-0.5 C rate till a voltage of 3.6-3.8 V per cell is reached, followed by constant voltage for a specified time. This may be followed by a float voltage of 3.5 to 3.6 V indefinitely to allow for equalization via the BMS. Charge maintenance is less important for lithium-ion cells which do not suffer from appreciable capacity loss, and are not damaged by extended periods at less than full charge.

It is possible to charge a considerable series-connected string of batteries with a single charger. It is much more problematic to charge parallel strings of batteries from a single charger.

It is almost inevitable that one string or the other will take more of the current. If one string has a failed cell, it will take much more if shorted, and much less if high resistance develops. The failure is thus enhanced instead of counteracted, and in the extreme it is possible to destroy one or both strings by over or under charging. It is desirable to use a separate charger for each parallel string.

Switching chargers can also be used as DC to DC converters. If one supplies a DC input, typically the pack voltage, it passes through one side of the input bridge and the rest of the circuit functions normally to produce the designed output, typically 12 V DC.

The Battery Management System

Ideally this is implemented for each individual cell with provision for equalization of charge during the charging cycle. A variety of BMS are available ranging from quite sophisticated systems with monitoring, as provided in expensive batteries such as those from Valence Technologies, to relatively primitive variants from the low-cost Chinese suppliers. Typically they limit the voltage range per cell from 2.0 or 2.5 V up to 3.6-3.8 V with a mean operating voltage of 3.2V.

Lead-acid batteries and all aqueous electrolyte systems including nickel cadmium and nickel-metal hydride are tolerant of overcharge because it simply decomposes some of the water in the electrolyte to hydrogen and oxygen. This gas is vented in flooded batteries (which may pose a hazard). **Proper venting and ventilation of flooded lead-acid batteries is essential**. Zero maintenance batteries such as gel cells and Absorbed Glass Mat (AGM) lead-acid cells recombine the hydrogen and oxygen internally. AGM batteries have vents to dispose of excess gas, denoted as Valve Regulated Lead- Acid (VRLA) batteries. The equalization process simply overcharges the strong cells in a series string in order to charge the weaker ones fully.

Lithium-ion batteries cannot tolerate overcharging because their organic electrolyte cannot decompose reversibly like water, it is simply destroyed, and the electrodes along with it. They will not tolerate over discharge either, and suffer irreversible damage at cell voltages below 2.5 V. **It is therefore critical that these very expensive lithium-ion batteries be protected both during charge and discharge by a battery management system that can supervise and control the operation of each cell.** This is particularly important with lithium-ion batteries having cobalt or nickel-based cathodes and a nominal cell voltage of 3.7 V. These cells are the ones responsible for the hazardous reputation of lithium-ion batteries in that they combine a strong oxidizer deposited on an aluminum foil cathode in the presence of an organic (flammable) electrolyte and a lot of stored electrical energy. They will burn fiercely if something goes wrong, and a number of vehicles have been destroyed this way.

The now common large format lithium-ion cells based on lithium iron phosphate with a nominal cell voltage of 3.2 V are less capable but much safer. They can still be destroyed by over charge or over discharge and require battery management systems. Valence incorporates much of the BMS functionality into the batteries in a very elegant package. Other manufacturers and independents offer separate BMS with various levels of sophistication. They all monitor individual cell voltages and should have an output that can function to shut off the charger if a cell is becoming overcharged and to shut off the pack if it is being overdischarged.

The more elegant systems monitor cell temperatures with similar protection against overheating during charging and discharging. Fortunately lithium-ion batteries do not heat up as much on charging as nickel metal hydride, and the phosphate batteries hardly heat up at all, requiring no cooling, a major simplification. Still the temperature sensing function is valuable in guarding against failed cells that do overheat.

The better BMS also have provision for cell equalization to keep everything uniform and at top capacity. Some do this by wasting power bypassing the high cells. More sophisticated systems transfer energy from the high cells to the low ones. However they work, this is a very valuable feature because even identical cells will in time drift out of equality, and this has bad effects on pack capacity and life.

For the BMS to function there must be a contactor in the line between the charger and the battery pack or some means of current control in the charger itself. This contactor can be much smaller than the main motor contactors, being rated only for the constant current capacity of the charger. It is wired to the relay, typically a MOSFET, in the BMS controlling the charging process. There may be a similar relay, mechanical or electronic, to shut off one of the main motor contactors if the battery pack voltage or that of one of the cells gets too low. The Evnetics controller has a built in battery pack low voltage limit. Alternatively one can rely on pack voltage instrumentation to decide when to shut off the power. The drop in power at the end of capacity is unmistakable. Other safety features such as over temperature shut off or warning can be incorporated.

The more elegant BMS offer a state of charge meter output and warning signals for near exhaustion of the pack capacity. This is a very useful feature that is also available as a stand alone instrument measuring pack voltage and amp hours consumed. Other BMS systems will provide a display of the voltage of each cell in the pack to monitor their status.

The Control System

The problem to be solved by the control system is to intervene between the battery, which is a constant voltage source of DC electric energy, and the traction motor to provide the motor with a variable source of power in accordance with the needs of the vehicle. Two basic control systems are available, AC and DC.

Direct Current is the classical approach in use since Edison's time. It is convenient in that the battery delivers DC and series- wound traction motors, which are relatively inexpensive and effective, can use it. The control system can be as crude as a set of tap switches to tap the battery pack at increasing voltage levels. Smooth, incremental power delivery can be accomplished by the now universally used Pulse Width Modulation (PWM) approach made possible by inexpensive power semiconductors.

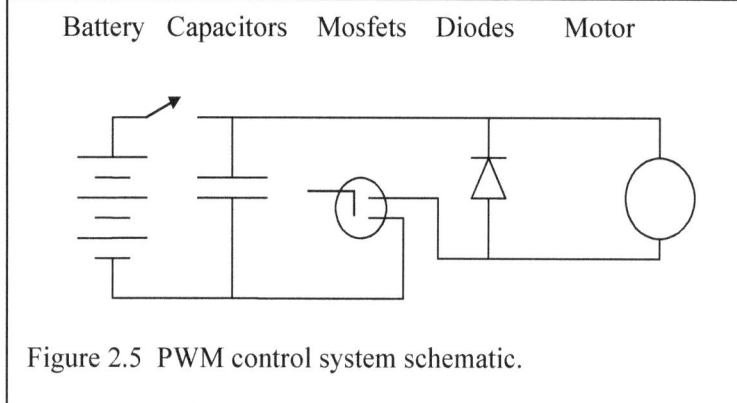

Figure 2.5 PWM control system schematic.

PWM functions by chopping the constant voltage battery supply voltage into a variable width square wave with semiconductors such as MOSFETs or IGBTs controlled by a timing circuit. The power electronic circuit is shown in Fig. 2.5 without the details of the timing circuit.

PWM effectively delivers a variable voltage to the motor. The current flow to the controller is smoothed into a constant DC current by parallel capacitors in the controller input, and the current flow to the motor is smoothed into a constant output by a parallel bank of "freewheeling " diodes in the output. The controller functions as a current transformer in which the output current can be larger than the input current in proportion to ratio of the total time to the "on" time, and the effective mean output voltage is lower than the input voltage by the ratio of "on" time to total time. Typically the controller limits the current to the motor to a value that will not damage the freewheeling diodes. The level is software adjustable on the Zilla and the newer Evnetics Soliton controllers. The current from the battery, which is more critical, can be limited by an adjustable potentiometer on Curtis controllers and by software for Zilla and Evnetics controllers.

AC control systems function by changing the battery DC voltage into a variable frequency Alternating Current to drive an AC motor at the desired variable speed. They do this again with power semiconductors functioning in this case as a variable frequency inverter. Typically, the motor is either a low-cost three phase "squirrel cage" induction motor or more recently a synchronous motor with high strength neodymium-boron-iron permanent magnets in the rotor and the variable frequency windings in the stator. The advantage is use of a low-cost, high-efficiency, high-speed AC motor compared to a DC motor limited by the commutator and brushes to a somewhat lower efficiency and speed. A further advantage is that the battery charger and a 60 Hertz inverter function for Vehicle-to Grid (V2G) integration can be built into the controller to achieve multiple functionality to offset the cost of the expensive power electronics package. The circuit is shown schematically in Figure 2.6. The circuit functions by alternately connecting the three phases of the wye connected motor to the positive and negative poles of the battery. Again there are capacitors to smooth the current flow and diodes to provide a current path when the semiconductors switch off. The semiconductors may be power MOSFETS or better, IGBITS (Insulated Gate Bipolar Transistors). They are in turn controlled by a frequency generating circuit (not shown) which provides gate signals at a variable frequency chosen by the vehicle operator to drive the motor at the desired speed.

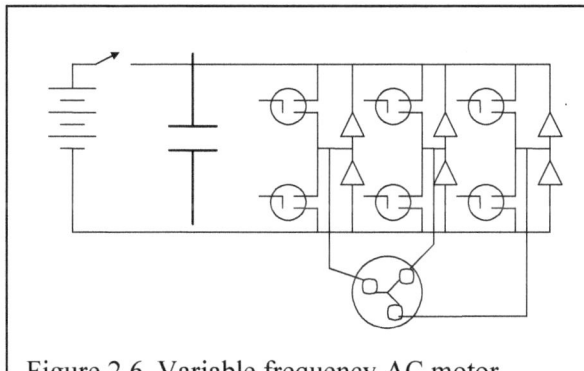

Figure 2.6. Variable frequency AC motor control circuit showing three phase inverter.

One can see that the same circuitry can be driven at 60 Hz to provide AC power back to the grid. Similarly one or all of the three phases may be connected to draw power from the grid to charge the battery. While a high power AC drive may cost two to four times the cost of a DC drive, this added functionality could pay off ultimately if the vehicle owner is reimbursed for the value of the distributed energy storage represented by the battery, as discussed below under V2G.

The advantage of a high-speed, low-cost motor can also be realized by the use of a brushless DC motor in which the commutation is accomplished optically or with a magnetic Hall sensor on the motor. This is a variant of the externally commutated AC system described above with the same type of power electronics. It has been used in the hybrid Honda Insight. Yet another variant is the switched reluctance motor with the simplest rotor configuration of all and an on-board commutated set of stator windings.

Motors

As mentioned above the two types of traction motors in use today are DC motors with commutators and three phase AC motors.

Series wound DC motors are constructed with a laminated iron armature containing conductor bars in slots, and, as the name implies, stationary field coils with the windings in series

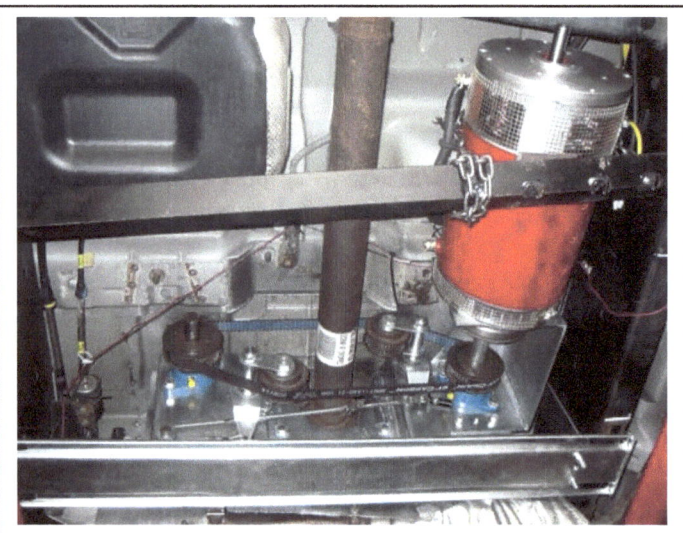

Figure 2.7 Series wound DC motor being installed.

with the armature. The armature is electrically connected to the outside world through carbon brushes which bear on copper commutator bars brazed to the conductors. As the armature rotates, the conductors are connected to the power source, first in one direction and then in the other, to provide a constant torque as the armature rotates through the magnetic field of the stator. Typically the motor will have two pairs of brushes and four poles for better utilization of the copper and iron, so the armature windings see an AC current at a frequency twice the rotational speed of the motor.

As the conductors pass through the stator field a voltage is induced in them which opposes the flow of current. This voltage is proportional to the field strength times the speed of the rotor. If the battery voltage, as established by the controller, is greater than this voltage, current will flow in the forward direction, a torque is imposed on the rotor and the motor generates mechanical power nearly equal to the electrical power represented by the voltage times the current.

If the voltage is less than the induced voltage, current will flow backwards, and the motor will act as a generator. This feature is used for regenerative braking in which the kinetic energy of the vehicle is converted into electric energy and returned to the battery. A related factor is that to achieve higher motor speed it is necessary to impose a higher battery voltage. High speed means high power from a given amount of copper and iron in the motor, and there is an incentive to match the battery pack to the maximum speed that the motor can withstand. Both ends of the armature and of the field are typically brought out to terminals on the outside of the motor so that the field can be reversed relative to the armature, thus reversing the direction of rotation of the motor electrically. Conventionally the armature is connected to the positive battery terminal.

Permanent magnet DC motors have the same armature construction but use permanent magnets to provide the field. They can be reversed electrically by reversing the connections to the armature. They are typically small horsepower motors used in appliances and the like, but there is a subgroup of radially wound, axial field permag motors originally sold as Etek motors now Lynch, Perm, Agni and other makes, which are used for light vehicle propulsion. They are rated at around 10 horsepower and are very light at 25-30 pounds. They are not rated at exceptionally high rpm typically around 3600. They are wired as 8 pole machines.

Figure 2.8 Agni 155R 27 cm. axial flux motor installed.

Agni has recently come out with a 270 mm (10.6") diameter 127mm (5") long motor rated at 23 kW (30.8 HP) at 120 V and 3600 RPM, which is a 12 pole design known as the 155 R because it has 155 radial conductors. It is the preferred candidate for a short twin installation where space is limited as described in Chapter 1 and shown in Figure 2.8. It requires almost equal sprocket diameters on the motor and the drive shaft because of its low speed resulting in a 1.26:1 speed ratio. Because of the very thin brushes of axial flux motors the brush life has proven somewhat problematic, as discussed in Chapter 1. The A155 weighs a mere 20 kg saving 250 lb in a twin installation relative to series-wound motors.

When ordering AGNI motors it is important to specify clockwise rotation facing the drive end of the motor to match the clockwise rotation of the IC engine from the front. For series-wound motors you will need to adjust the brushes yourself by removing the screws holding the brush assembly to the body of the motor and rotating the brush assembly to advance the brushes against the motor rotation to the furthest set of screw holes.

AC motors have the advantage that the necessary AC current to cause rotation is available without the need for a commutator, which is a troublesome component. The two AC motor types used in most applications are induction motors and synchronous motors.

Induction motors are called squirrel cage motors because the rotor is a stack of radial iron laminations with axial conductor bars which are short circuited at each end by a welded or riveted ring, looking like a barrel-shaped cage of conductors. Stationary windings are distributed around the circumference of the stator with laminated iron cores to provide an alternating magnetic field. The three phase motor is particularly simple with three coils. As the three coils are successively excited by the individual phases of the power supply they create a magnetic field that rotates. To reverse the direction of rotation, reverse any two of the three connections. When the rotor is exposed to this rotating field, current flows in the squirrel cage conductors to create a field opposing the applied field, and the interaction of the induced currents and the field creates a torque to start the motor. Inrush currents can be very high, since the rotor acts as a shorted transformer until its speed builds up. At no load the rotor will accelerate to nearly the speed of the imposed field, 3600 rpm for a two-pole motor at 60 Hz. Under load, the rotor speed is reduced and the current increases, the input electric power matching the mechanical load on the motor. As traction motors, induction motors suffer from low starting torque and poor performance at low speeds and low frequencies. This can be mitigated by increasing the frequency which speeds up the motor and makes it proportionately smaller and lighter for a given power while requiring more speed reduction to drive the load.

The AC propulsion motor shown in figure 2.9 is a four pole induction motor with a maximum speed of 12000 rpm at 400 Hz, the standard military AC frequency for aircraft and ships. A single AC Propulsion AC-150 motor could easily provide the required power for a

47

pickup truck conversion. It is capable of delivering 150 kW maximum power and 50 kW (67 HP) continuously at 7000-8000 RPM from a 336-360 V DC source. This corresponds to 225 Nm (170 lb-ft) of maximum torque and 73 Nm (54 lb-ft) continuous torque from 0 to 8000 RPM. The continuous power rating is almost exactly twice that of the 9" series wound DC motors and will not overpower 8 mm pitch Poly Chains which are rated at 52 HP each at 5500 RPM for a total of 104 HP with 34 tooth sprockets. Use of 40 tooth motor sprockets can raise this to 123 HP.

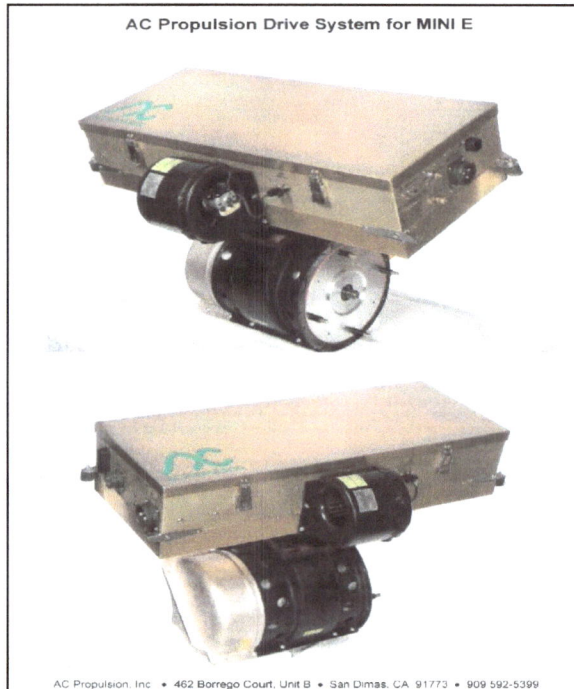

Figure 2.9 Synchronous high speed AC motor and Power Electronics Unit by AC Propulsion.

The AC Propulsion drive is much more expensive that the other alternatives. The advantages are: 1/3 the weight, saving 200 lb relative to series wound DC motors, unlimited speeds, regenerative braking, and built in grid integration with 20 kW of battery charging and inverter output power capability. In a V2G environment the DC and AC Propulsion alternatives may be essentially equivalent with the AC Propulsion option clearly more sophisticated and with more potential. This provides strong incentive to demonstrate a single motor drive that can utilize it.

An alternative for a single motor drive is the 11" diameter series-wound Warp 11 from Netgain. It is capable of approximately twice the torque and power of the Warp 9s used in the Big Twin installation.

Synchronous AC motors have the same windings on the stator as induction motors, but the rotor contains either wound or permanent magnets. The advantage, especially with permanent magnets, is a simple rotor structure which can achieve very high speeds and thus high power-to-weight ratios. Modern extreme strength neodymium-boron-iron magnets have made possible a major advance in design of synchronous AC and axial field DC motors. The other advantage is that while the synchronous motor is poorer at starting with a fixed frequency than an induction motor, it is better at running with variable frequencies. The most elegant electric vehicle drives now use the variable frequency principle with synchronous motors turning at speeds over 10,000 rpm and extensive gear reduction to the final drive. Very high power-to-weight ratios have been achieved, typically twice those of DC motors.

Finally there is a type of AC motor called the Switched Reluctance motor, so called because the magnetic path of lowest reluctance (highest permeability) is switched from pole to pole of the stator dragging the rotor with it. Typically such a motor will have a four pole steel rotor and a six pole wound stator. The rotor is simpler than an induction motor because the magnetic field through it is always in the same direction and there are no windings or electrical connections. The rotor can be hogged out of a single piece of steel. Demands on the control circuit are higher than for variable frequency induction motors because the current flow to the windings must be accurately timed relative to rotation and the inductance of the windings is variable. The trend toward higher costs for complex mechanical structures and lower costs and better performance for power electronics may favor this type of motor in the future.

48

The Drive

The Twin Motor Drive

The drive unit connects the motor to the vehicle drive shaft and it has been the major innovation of this development program for supplemental PHEVs, as described in Chapter 1.

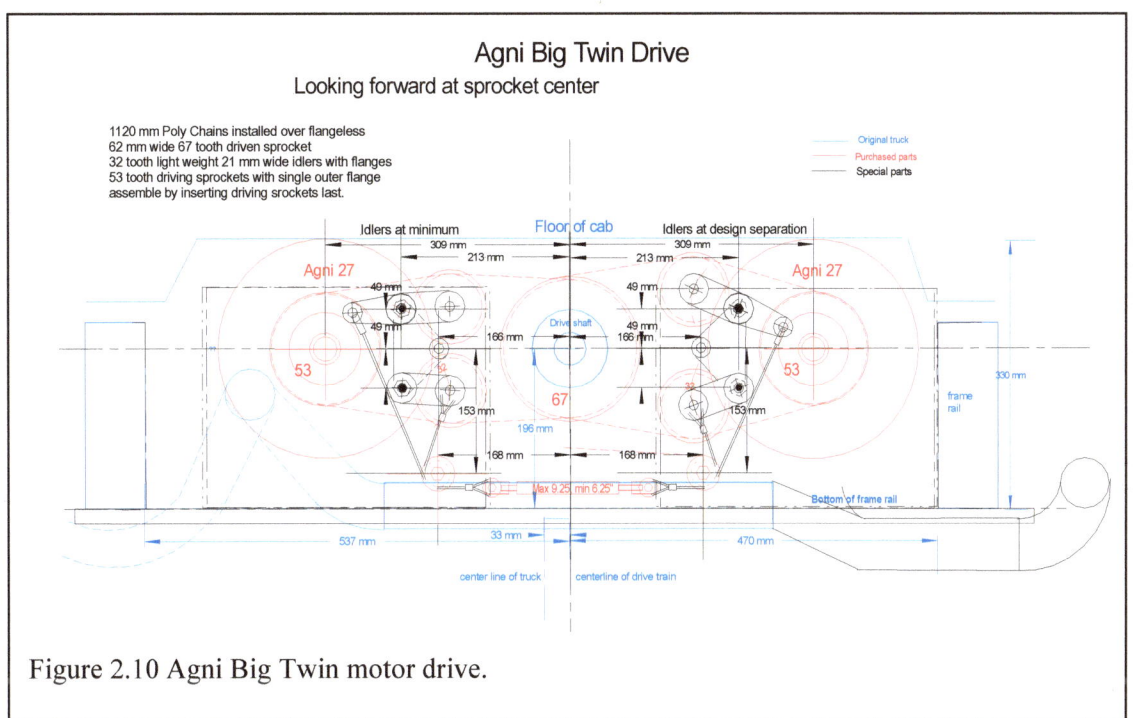

Figure 2.10 Agni Big Twin motor drive.

Where there is adequate space under the vehicle the twin-motor drive with Agni 155-R 270 mm diameter axial flux motors shown in Figure 2.10 and in **Drawings 8-12** is preferred. The AGNI being only 125 mm long is able to fit under the sheet metal better than the longer series-wound motors. This also helps solve the problem that some trucks have larger gas tanks, four wheel drive transfer cases or shorter wheelbases that don't provide enough room on the left side for a series-wound motor.

The same drive can be used with 9" series-wound DC traction motors or with the EVS AC-50 AC motors, The only major change is that both motor sprockets must then be lowered by 51 mm relative to the drive shaft center line to allow both motors to clear the truck underbody. (With a single motor only the motor side needs to be low, the idler side is correspondingly high). Lowering the motors requires a pair of smooth faced idlers bearing on the back side of the drive chains to allow the idler brackets to be positioned on the same level as the drive shaft sprocket and exert balanced forces on it. While back side idlers of the appropriate diameter are allowable with Poly Chains, they are a source of wear, as well as being an extra component. This is offset by the fact that the back side idlers provide a convenient way to install the chains over the flanged sprockets by installing the idlers last.

The drive sprockets of the twin drive are mounted directly on the shafts of the two motors for simplicity and because it is necessary to have the motors as far forward as possible to clear the gas tank. Because the motor shafts are short, the motors are offset axially to match the offset between the two driven sprockets and to minimize the overhang of the sprockets.

Gates Rubber Co. Poly Chains have been chosen for this service because they have the power transmitting capability of steel chain, but require no lubrication, can be run in a wet-dusty environment and are silent when properly adjusted. Goodyear Eagle and other brands are available. They feature deep rounded teeth on a polymer belt reinforced with carbon fibers for strength and stiffness. Belts such as these have become standard on motorcycles because of higher power transmitting capability higher speed capability, low noise and unlubricated compatibility with a dirty environment.

The chains must be tight to transmit the considerable torque amounting to over 28 ft lb each. At the same time the drive has to be compliant to allow the drive shaft to move freely in both radial directions to accommodate vibration and movement under load of the IC power plant. This is accomplished by the arrangement of idler sprockets shown in Figure 2.10. The driving chains are tensioned by opposite pairs of idlers linked by cables and a turnbuckle. As the turnbuckle is tightened, the chains are tightened on the sprockets, and the idlers are spaced evenly top to bottom to clear the support posts between them. The chains should be tightened so that a ten pound load at the center of one of the runs deflects the chain about one chain thickness (1/4 inch).

Despite the considerable tension on the chains, the drive shaft sprockets can still move up and down because the chain runs leading to the big sprockets are almost horizontal, and vertical movement of the drive shaft causes little change in tension.

In the horizontal direction one can see that drive shaft movement to the left causes the two right idler sprockets to come together and the two left sprockets to move apart, and vice versa. Again, even though there is a lot of tension on the chains, there is little resistance to movement. To make this system function at high frequencies of vibration it is desirable to reduce the inertia of the idler system. This is the reason for making light weight idler support brackets and boring out the bolts holding the idlers. The idler sprockets themselves weigh 2.3 lb with the hollowed out bolt and could benefit from further lightening. To equalize the chain tension top to bottom a torsion spring is used on the lower idlers to support their weight and that of the linked upper idler on the other side. A counter weight on the lower idler brackets would also provide compensation but increase the inertia undesirably. This compliant drive concept is the subject of a pending U.S. Patent[xvi].

The tensioning cable arrangement uses fixed loops of cable on each side and a single turnbuckle because the objective is a simple side to side motion of the cables. In earlier versions with two turnbuckles and cables linking upper right and lower left idlers there is a twisting vibration mode with the upper right and lower left idlers moving out while the lower right and upper left move in and vice versa, which is undesirable.

A light twin drive based on 7" series wound motors has been built and the layout and parts are shown in **Drawings 27-29.** In this case the offset relative to the drive shaft center line is minimal and the back side idlers were not used. The light twin drive is really inadequate for a full size pickup, and is shown for completeness and because it may be of use in lighter vehicle conversions such as early model cars and light pickups.

The Single Motor Drive

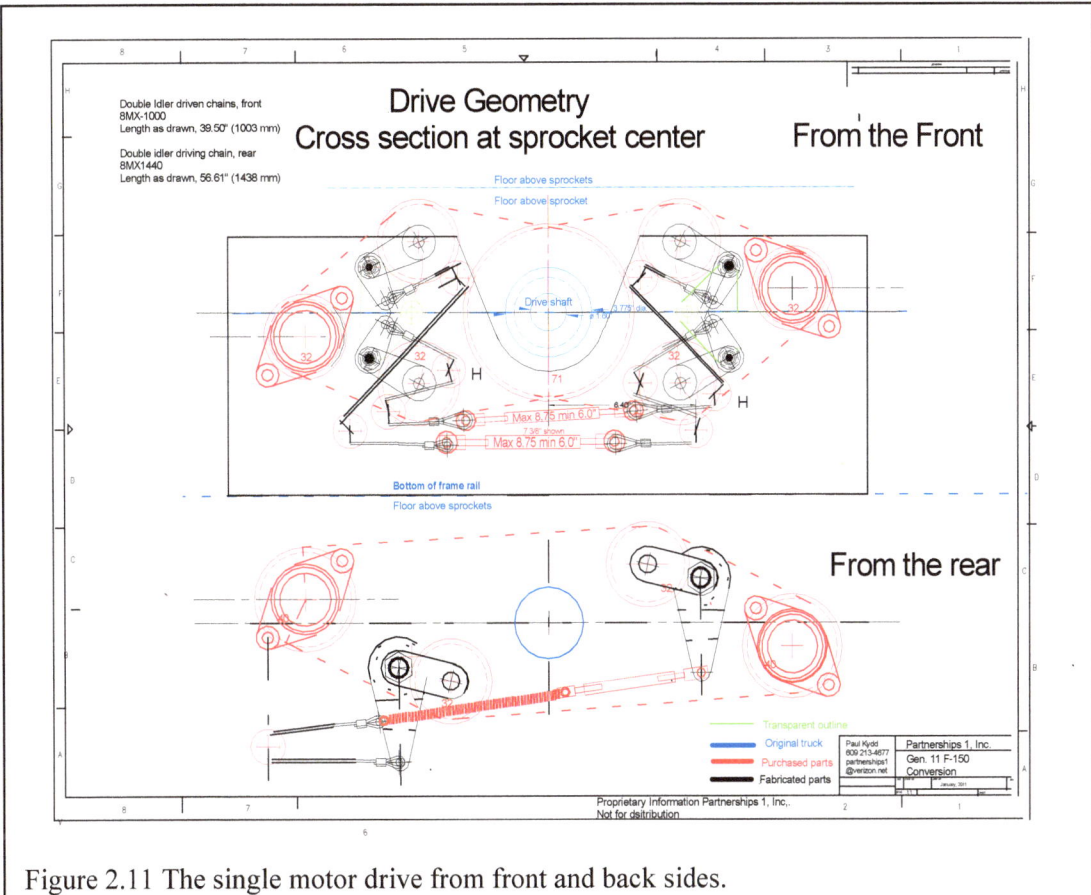

Figure 2.11 The single motor drive from front and back sides.

A great deal of effort has been directed at perfecting the single motor drive shown in **Drawings 18-26** and in Figures 1.3, 1.4 and 1.5, because of the desire to use AC motors like the AC Propulsion system shown above, and to drive trucks with limited space under the cab. This creates a serious problem in that it is necessary to take torque from a single electric motor and apply it to both sides of the drive shaft to provide balanced forces and a pure torque on the shaft. This has been done by using a rear-mounted cross chain to drive two shafts from one rear-mounted motor as shown in Figure 2.11 above.

The front side of the drive is very similar to the twin drive described earlier. Two ball-bearing-mounted shafts drive sprockets on two front-mounted driving chains which drive the vehicle drive shaft. The two shafts turn at up to 5740 RPM with 34 tooth sprockets.

The rear side of the drive comprises two 40-tooth sprockets on the two shafts and a Poly Chain transferring power from the right hand shaft to the left one. The right hand shaft is driven by the single motor which is spaced back from the drive by a distance piece to clear an obstruction on the floor of the truck cab, and to provide space for a shaft coupling and a device to synchronize the front and rear sides.

Once the front side chains are properly adjusted and tensioned, the motor side chain can be tightened and locked in position. This chain is tensioned by two idler sprockets to allow it to be accurately synchronized with the two driving chains on the front side which are rigidly locked together by the main sprockets on the drive shaft. A spring and cable are used to tension the cross chain idlers while allowing for stretch and wear of the chain. It is critical that the cross chain idlers be restrained from yielding under torque from the motor or from the drive shaft windmilling the motor. This can be accomplished by overrunning clutches in the cross chain idler brackets or by simple levers engaging racks mounted on the base plate as shown in **Drawing 22**. These restraints allow the idlers to tighten the cross chain but not to loosen it. Strictly speaking the cross chain idler brackets should be spring compensated for weight like the driving chains, but the force on them is so low compared to the tensioning force that this is not necessary.

As mentioned above, there is the potential for noise and vibration if the cross chain gets out of synch with the driving chains. This is inevitable due to stretch and wear of the chains no matter how carefully adjusted initially. Even the self adjusters described above do not entirely eliminate it. To counter this problem a differential gear arrangement, shown in **Drawing 23**, has been designed to cure the problem, but it has not yet been tested.

Also not tested are single-motor conversions using the Netgain Warp 11" motor or the AC Propulsion AC-150, which are required to drive the truck at an acceptable speed continuously. As mentioned in Chapter 1, the Advanced DC FB1-4001 and Warp 9" motors are really not adequate for every-day use in single motor drives, being overloaded by a factor of two. The larger motors have the continuous duty rating for this service and **Drawings 19, 20, 23 and 25,** show these motors and other parts consistent with them. While these drawings are correct based on the information available on these motors, it should be realized that these conversions have not been done, and there is no assurance that the parts will in fact fit or that the conversions will have the expected performance.

I have elected not to pursue the single motor drive at the present time because it has proven to be complex and expensive to build and unreliable in operation. There is no current V2G infrastructure which can justify the cost of the AC Propulsion system, and the system seems to be unavailable to individual converters anyway. As a result I am recommending the twin motor approach which offers the ability to power a full-size pickup within the continuous-duty rating of available compact, low-cost AC and DC motors and a much simpler drive that has proven reliable. There is an additional advantage of the twin DC motor arrangement in that the motors can be wired in series so that the same current passes through each motor guaranteeing equal torque and an automatically balanced drive. Since balanced forces are a critical objective of this type of conversion, this is an important consideration.

Twelve Volt Power

The electric drive controls are powered by a plug-in adapter to one of the cigarette lighter 12 V sources on the dash. This provides 12 V power from the truck's electric system to operate the PLC, the controller and its cooling fan, and the contactors in the control compartment. A switched 12 V receptacle that is on only when the IC engine is running is a desirable source, which is automatically turned off when not in use to avoid accidentally running down the vehicle starting battery. If no switched receptacles are available, a switched accessory plug is acceptable for conscientious drivers who remember to turn it off, but a better solution is to build the shut off function into the PLC by adding a connection to the ignition system.

The Electric Accelerator

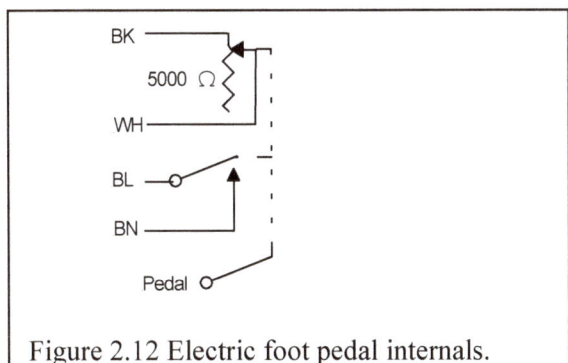

Figure 2.12 Electric foot pedal internals.

The electric drive is controlled by a separate foot pedal immediately in line with the IC engine accelerator pedal, as shown in Figure 1.1 such that the driver's foot normally encounters the electric accelerator first. This enables the driver to automatically use as much power as the electric system can deliver and supplement it with IC power as needed by pressing further to activate the IC accelerator/throttle. The pedal assembly from Curtis is linked to a 0-5K Ω potentiometer for the PWM controller input black and white pair and a microswitch for the secondary contactor blue and brown wires as shown in Figure 2.12.

The PLC

For those conversions not using the Evnetics controller, which has built in logic to interlock electric and IC operation and inhibit the electric drive when in reverse, when the brakes are applied or when the battery is low, a Programmable Logic Controller (PLC) located on the transmission hump can perform these functions as shown in Fig 2.13. The PLC has a display to show the status of the various inputs and outputs. Certain operating modes are not desirable with the combined IC and electric propulsion systems, and the PLC is used as an interlock to avoid them.

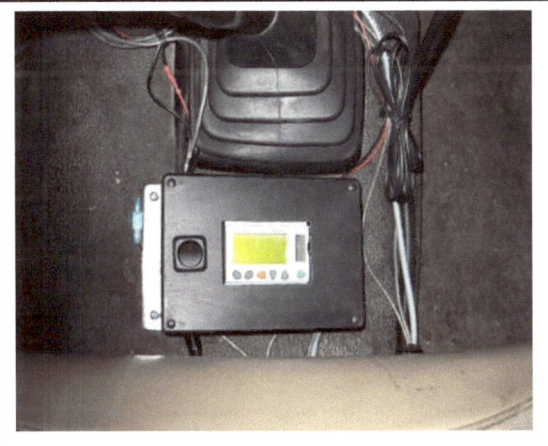

Figure 2.13 The Crouzet Millenium 3 Programmable Logic Controller.

The most obvious conflict is between the electric system which only operates in the forward direction and the transmission which has to reverse. To eliminate the conflict a +12 V signal is taken from the reverse light switch on the transmission to one of the PLC digital inputs.

With manual transmissions it is necessary to lock out the electric drive when the clutch is depressed to allow the engine to be started and the IC throttle to be operated without the electric drive. This is accomplished by taking a +12 V signal from the clutch switch, if there is one, or by putting a switch on the clutch to provide a 12 V signal to another of the PLC digital inputs. This has the consequence that every time the clutch is depressed to shift gears it is necessary to double pump the electric throttle to reset the Curtis controller which has an interlock preventing starting under load. This is annoying and renders application of the electric drive system to manual transmissions much less pleasant to drive than with automatics.

For automatic transmission vehicles a signal is taken from the transmission which normally allows one to start the IC engine only in Park or Neutral and used to allow the PLC to lock out the electric system in Park and Neutral to permit IC engine starting without interference from the electric drive. A switch on the PLC housing permits overriding this lockout when the vehicle is rolling to allow the driver to shift the IC engine into neutral while driving on electric power. This prevents the electric drive having to drive the IC engine as well as the vehicle, increases speed and saves electric energy. Vehicles with free wheeling in overdrive do not need this feature.

The PLC can also prevent simultaneous application of electric power and brakes by taking a +12 V signal from the brake light, but this situation almost never actually happens with a skilled driver.

Finally, the PLC monitors the battery voltage and shuts down the system when the battery is exhausted. Typically with heavy current draws, the battery voltage begins to sag due to limited diffusion of the ionic charge carriers into the electrodes while the battery still has significant capacity. Shutting the electric system down when this voltage drop begins to be significant is a safety feature which enables the driver to use all the battery capacity that is safe, but end electric operation while there is still approximately 10% of capacity remaining, to ensure against complete exhaustion and irreversible damage of any of the cells. The PLC performs this function by monitoring battery pack voltage through an opto-coupler and shuts down the system when the voltage falls to a preset level. A delay is built in to prevent shut downs from momentary voltage sags due to acceleration. Following shutdown the PLC can allow restart when the pack voltage recovers, or it can be latched in the off position until the battery pack is recharged and the pack voltage increases significantly.

Figure 2.14 Instrument cluster showing output current to the motor, battery voltage, and input current to the controller (battery output).

The Instrument Package

The instrument package shown in Figure 2.14 is located on the dash directly in front of the driver providing easy reference to the battery pack state of charge and the current draw. The former is provided by a volt meter or a state-of-charge meter with LEDs showing the battery capacity remaining. The current is read directly on an ammeter. Shunts are located in the truck bed electronics enclosure and light gauge cabling brings the voltage and current signals through the enclosures on the motors, the PLC enclosure on the transmission hump and through plastic loom up onto the dashboard. Both battery draw and motor input currents can be monitored. The instruments have internal lighting for driving at night.

What You Can Expect

Facilities Needed

Because the objective of the supplemental hybrid concept is minimum change and minimum cost, special tools and equipment necessary to do the conversion are kept to a minimum. In particular, modifications to the original vehicle are limited to rerouting the exhaust, removing two sheet metal brackets under the seats and drilling a pattern of holes in the vibration damper ring boss on the front universal joint yoke of the drive shaft. No precision machining or disassembly of the vehicle is required. The entire conversion can be done without opening the hood.

Modification of the exhaust is a specialized task requiring welding and metal forming skills best left to a professional. There are well equipped muffler shops in almost every city which can do the modification at minimal cost. I recommend having this work done professionally, especially since exhaust leaks can be dangerous to your health.

Installation of the battery pack is next. The mounts should be installed first and the cells later or you will require some means to pick up the 50 Volt modules weighing 100 lb each or more and putting them in the vehicle. Mounting them requires drilling through the truck bed or van floor and securing them with self tapping cap screws. It is necessary to drill through the side of the truck bed with a 2 ½" hole saw for the 54 inch length of 2" electrical conduit carrying the current and control and instrument wiring down to the motor. All of this work can be done with hand tools.

Installation of the electronic enclosure is next. This is prewired, installed over the battery pack and fastened into place. It requires no special tools or equipment.

The next task is to install the main sprocket assembly on the drive shaft. This involves removing the drive shaft, drilling and tapping the vibration damper ring, installing the main sprocket assembly with the Poly Chains looped over it and reinstaling the drive shaft. If you plan to balance the modified drive shaft, which is a good thing to do, allow for two sessions to remove and modify, and then reinstall. During balancing you can also check that the sprocket is running true on the front yoke of the U-joint.

Figure 2.15 Raising the electric drive into place with a transmission jack.

Installation of the electric drive itself involves working under the vehicle and lifting the 2-400 lb drive unit into place. While it is possible that this can be done with the vehicle jacked up, it will be much more feasible and much safer to do this on a lift with a transmission jack to raise the drive into place as shown in Figure 2.15. The unit needs to be raised into place, aligned with the main sprocket, and fastened into place with clips bearing on the frame rails of the vehicle. The motor electrical enclosure boxes and the rear support channel are installed, and the rear end of the motors supported, if needed. The Poly Chains are then installed and tightened.

The connections to the reverse switch, and park-neutral starting switch on the transmission and to the brake light switch are made and fed into the motor electronic enclosures. The wiring from the control system in the truck bed electronic enclosure is fed down through the 2" conduit, which is then connected to the motor electrical enclosures, followed by connecting the high current wiring to the motors and running the instrument and control wiring up into the cab through an existing hole in the truck cab floor. This completes the work under the vehicle.

The final task is to install the electric accelerator with two sheet metal screws, the PLC with four screws and the instruments with double-sided tape and complete the wiring in the cab.

Time Spent

All of the times estimated below assume that the major modules of the PHEV system have been preassembled to the maximum degree possible, which is highly desirable to minimize the time that the truck is immobilized and tying up an expensive lift. The times are for persons with reasonable skill at mechanical and electrical work but no experience in this conversion. The job will be much easier, faster and safer with two people. Persons working alone or with less background should allow twice as much time at least. Glitches due to parts that don't fit, etc. can multiply the time drastically, so try to ensure that everything is OK before beginning assembly. Professionals working with pretested components on a single model of truck should be able to cut these times by 50%.

Assuming that all the major components are available, the following times will be required to do the **preassembly**:

Battery modules-------One to two days with Valence batteries, lead-acid batteries or Manzanita reg. decks. Three to four days with individual cells and separately wired BMS.
Electronic control enclosure------one and a half days
Drive, motors and motor electrical enclosures ------one day
PLC ----one day
Instrument wiring----half a day
Count on a week of hard work to get ready for the installation.

Time required for **assembly:**
Allow at least a day to install the battery modules and the electronic control compartment, much of which will be locating and drilling the holes and working both above and under the truck to fasten the battery modules and thread the 2" conduit and wiring.

Installing the drive under the truck should take about another one to two days including modifying the drive shaft, installing the drive, installing and adjusting the Poly Chains and connecting up the wiring.

Final installation of the electric accelerator, PLC and instruments should take another day with shakedown, test runs and fixes to follow, total installation--roughly another week.

Cost

As mentioned above, the battery is the single greatest cost, amounting to approximately $2400 for a set of twelve Trojan 1275 flooded lead acid batteries for a minimum 144 V installation. Lithium cells will cost approximately $8,000 for a 100 Amp hour 205 V set, and proportionately more for higher storage capacity. You would be wise to order5 % more cells than you need to make up for early failures or accept the fact that your pack will be somewhat under powered. The BMS will add $2400 or more, and a set of Reg. decks another $2000 plus the display at $400, total cost, $13,000. A set of sixteen Valence group 24 110 Ah batteries, plus their

BMS will run close to $18,000, but the difference is well worth paying for added reliability in our experience..

The battery boxes, control compartment, drive components, and PLC/Instruments at current prices total approximately $14,000, comprising the following major components:

		Single Motor		Twin Motors	
		DC	AC Propulsion	DC	AC
Battery mounts		$400	400	400	400
Charger		$2000		2000	2000
Controller		$2000		2000	
Contactors & 2/0 wiring		$600	600	600	600
Control enclosure		$500	500	500	500
Motor(s)	DC	$2600*		4000***	
	AC		$32500**		9200****
Drive components		$4200	$4200	3500	3500
Motor electrical boxes		$200		400	400
PLC		$300		300	300
Instruments		$400	2400	400	400
Misc. wiring & fastenings		$300	300	300	300
Total	DC	$13500		$14200	
	AC		$40900		$17400

*Netgain Warp 11. ** AC propulsion 2009 price for one drive. ***Two Netgain or Advanced DC motors. (AGNI motors $500 additional).   ****Two AC-50 motors and controllers.

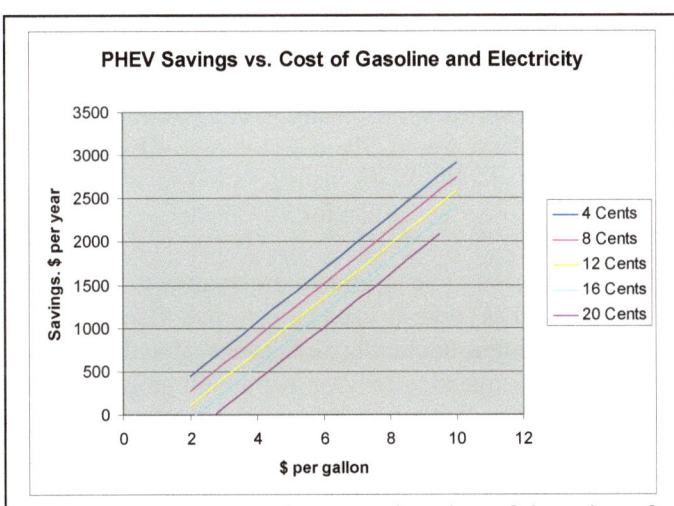

Figure 2.16 PHEV savings as a function of the price of gasoline in $ per gallon and electricity in cents per kWh.

The total for components is thus in the neighborhood of $30,000 for a minimum lithium system and $16,500 for a minimum lead-acid system. To which must be added the cost of exhaust modifications, roughly $200, and any other professional help including the use of a lift which might amount to $700 for the day of work under the truck modifying the drive shaft and installing the drive, say another $1000 total.

There is also the cost of a charging station at home which can run all the way from $250 worth of parts for a self-installed 240 V circuit to $2500 or more for a professionally installed, specialized station, plus roughly $1000 more for a separate service if you wish to implement V2G at more than 10 kW.

Savings

Most pickups are driven 12,000 to 20,000 miles per year. Assuming that the vehicle has a minimum 20 kWhr pack and is charged overnight and driven at least 30 miles per day winter and summer for 350 days per year (10,500miles), the annual saving is shown in Figure 2.16, based on 17 miles per gallon before conversion, 50% fuel saving due to conversion and 0.40 kWh per mile average between local and highway driving. The nationwide average price for electricity is approximately 12 cents per kWhr. At this price PHEV ownership breaks even with $1.75/gallon gasoline and nets about $1000 annual saving at $5.00 per gallon. Lower electric prices and higher gas prices of course increase the savings, but they are relatively modest for a $16-30,000 investment.

In addition to serving as insurance against rising gas prices the conversion can protect against the loss of value of the vehicle itself due to high fuel prices. In the last oil price spike in 2008, pickup trucks typically lost something like 20% in resale value because of the high cost of refuelling. The same is likely to happen in future, and could represent a $3000 advantage for a conversion.

Implementing Vehicle-to-Home (V2H) and Vehicle-to-Grid (V2G) for Savings

Your conversion can play a role beyond saving gasoline in transportation. There are Vehicle-to-Home or (V2H) opportunities to integrate the massive storage battery you have invested in with your home and Vehicle-to-Grid or V2G opportunities for integration with the electric grid.

V2H

Vehicle-to-home integration can provide you with a backup source of power for critical home functions such as emergency lighting and refrigeration in case of power outages. The necessary interface is an inverter to convert DC power from the battery to AC power you can use. The technology has been developed for off-grid solar systems. Magnum Energy of Everett, WA, is an example. They offer MS PAE inverters rated at 4400 Watts that operate from 48 V battery banks to supply 120 V or 240 V single phase AC power which list for $2700 each.

These units also serve as chargers, and one way to configure the system is to buy four of them and allow them to charge your battery as well as serve as back up inverters. Each charger/inverter would be connected to each of the four nominally 50 V packs. You do need to disconnect the four packs which would have to be isolated to provide a common ground when charging because the Magnum inverters cannot be stacked on the DC side. They can be paralleled on the AC side.

This is a rather expensive way to go because you will also need:

Inverters	$10,800	A router	399
Switching for Parallel op.	800	Grand total	$15,460
An Enclosure	2469		
Extension boxes 2 @	329		
Back planes 2@	169		

The Magnum units disconnect from the line automatically in the event of a power failure preventing your battery supply from electrocuting line workers who think the line is dead. To take advantage of the emergency power you will need a separate sub panel for essential loads upstream of the inverters. This is a major installation, but it will give you access to the entire 20 kWh of energy in your battery pack which should keep a refrigerator-freezer running for two days or so, and provide enough power for well pumps or cooking ranges, even air conditioning. If you have a solar photovoltaic array the Magnum system will keep it working during a power outage, which, frustratingly, it otherwise will not do. However, the real advantage of the maximum installation is the possibility of V2G discussed below.

It would be possible to lower the cost of the installation, (and keep the advantage of emergency solar operation), by paralleling the 50 V battery packs and feeding one inverter for 4 kW of AC emergency power. The sub packs could be charged with individual golf cart chargers at about $300 each. It would then be necessary to shift back to series connection to run the vehicle. There are several suppliers of emergency power inverters like the Magnum, but they all operate at 48VDC because that is allowed by the National Electrical Code.

A more convenient but more risky option is to use a single high voltage inverter such as those used in uninterruptible power supplies at approximately $1000 to provide a few kW of 120/240V AC power direct from the 200-230 V DC of the battery pack. This by itself is enough for refrigeration, limited cooking, keeping the gas heating system running and some lighting. Your 20 kWhr battery will keep the house running for at least a day and more if you restrict high power loads like cooking and air conditioning. You do need to do the switching from the grid to battery backup manually, and you have to make sure that they cannot both be connected simultaneously. You also need to get the system permitted and inspected by your local municipality to keep your insurance in force and to stay legal.

Yet another option is to mount the high voltage inverter on the vehicle. In this case the emergency power is available wherever the vehicle is. This could be particularly valuable for tradesmen who need power at the job site and sportsmen who need TV when they are camping in the unspoiled wilderness.

V2G Opportunities

Vehicle-to-grid integration is a relatively new concept pioneered by AC Propulsion of San Dimas, CA, who have adapted their high end electric vehicle drive system to support it. The idea is to provide electric storage to the electric utility grid by aggregating individual vehicle batteries into a large enough capacity to bid on supply of "ancillary services" to the local Independent System Operator (ISO) who is responsible for grid operation and stability. These ancillary services are frequency regulation, spinning reserve and peaking power or demand management. The local ISO in New Jesey, (actually an RTO, Regional Transmission Operator), is PJM whose territory includes New Jersey, Pennsylvania, Maryland, Delaware and much else as far West as Chicago. North of us is the New York ISO and New England ISO. To the West is Midcontinent MISO. Texas has their own ERCOT, and California has CAL ISO.

The ISO is responsible for maintaining the frequency of the grid at exactly 60 Hz and in phase with the adjoining territories. This requires the ability to add or shed load very quickly in response to big circuit breakers opening or closing on the system. To do this the ISO solicits bids for power to be available at 4 seconds notice for each of the following 24 hours and accepts bids equal to approximately 1% of the total system capacity. The average price paid by PJM for this service is approximately $30 per megawatt of capacity per hour connected. Batteries are particularly good for frequency regulation because they can respond much faster than rotating machinery. The Federal Energy Regulatory Commission has issued Order 755 which mandates

ISOs to specially compensate fast response capability, and PJM has responded with Regulation D frequency response with 1 second notice and significantly higher value.

Your 20 kWhr battery with its four Magnum 4.4 kW inverter/chargers can deliver or consume 17.6 kW essentially instantaneously. This is worth something like $0.50 per hour when plugged in, or $4400 per year if plugged in all the time and bid successfully every day. Neither of those things is going to happen, but even if plugged in 90% of the time and bid successfully half the time, the purse is $2000, which will pay off the inverter/charger installation in seven and a half years.

The key is the aggregation of a number of vehicles into a large enough entity to bid for this revenue. Professor Willet Kempton of the University of Delaware has been a proponent of this concept and has published widely on it over the last decade. There is a very good book describing the concept "V2G 101" by Leonard J. Beck, 2009[xvii]. The technology has been demonstrated in Delaware with a group of eight "E-Boxes", Toyota Scions that were converted to electric vehicles with AC Propulsion drives and U Del. communication hardware and software by Auto Port, a Delaware conversion company.

A small number of vehicles could be used to demonstrate ancillary services which normally requires up to 100 kW of capacity because they were piggybacked on a one megawatt battery installation on the PJM system owned by AES, an international electric utility company. Pure battery storage is used in a few isolated utility locations now and is expected to grow. The advent of large amounts of solar and wind power will require additional storage because the wind doesn't always blow and the sun doesn't always shine, and they can't be controlled. Some controllable storage capacity will be needed to buffer the renewable supplies and keep the grid stable. The Danes and the Germans with larger percentages of renewable energy than the US are already encountering this problem. The answer is vehicles, and plenty of them. The vehicle batteries, even after they have begun to loose capacity and are no longer suitable for transportation, will still have value as stationary storage where maximum capacity per unit weight and size don't matter.

Achieving this ultimate, optimized 21st Century Personal Transportation Energy Supply system will require considerable innovation and investment in communications and utility regulation. A start has already been made by NRG, an innovative electric energy company headquartered in Princeton, NJ. They are offering energy supply contracts under a concept known as eVGo to electric vehicle owners in Texas, where electric utility regulation is minimal. NRG will supply the electricity for charging at so much a month, similar to a cell phone contract, and calculated on a basis to save the owner money relative to gasoline for a typical mileage driven. NRG thus gets all of the advantages of V2G and can supply power primarily at night when there is excess capacity and it is cheap. The owner gets a benefit relative to gasoline and the simplest possible interface with the supplier. NRG has licensed the University of Delaware technology as part of their planned introduction of automotive vehicle energy supply as a service named v2Go. Demonstration work is continuing with Mini-E vehicles from BMW.

This may be the model which will ultimately be adopted. It makes sense that the utilities, which deliver the power, do the demand management and all of the book keeping on the rates. They may elect not to do that, however, and there will be opportunities for other aggregators, even groups of vehicle owners, to step in and offer this service. PJM has lowered the threshold at which an aggregator can bid to supply regulation service to 100 kW. With a 17.6 kW DC system or the 20 kW AC Propulsion system it only takes an aggregation of 5 or 6 conversions to be a player. Partnerships One, LLC. has begun development of such a system known as Vehicle-Solar-Grid Integration.

Vehicle-Solar-Grid (VSG) Integration

VSG is a new vehicle-to-grid concept that is being tested at the recently built GridSTAR facility of Penn State University at the Philadelphia Navy Yard.[xviii] The advent of large numbers of electric vehicles will soon provide a major resource of distributed electric storage capacity in the U.S. Typically these vehicles are in use only a few hours a day representing an underutilized asset which can be plugged into the grid to provide ancillary services. Similarly there is an increasing number of solar photovoltaic installations with grid-tied inverters which are idle during the night time hours. VSG seeks to provide the key linking technology to combine these underutilized assets to provide distributed, dispatchable electric storage capability to the grid.

The schematic diagram Figure 2-17 shows VSG interface equipment which we call an EVPV, linking an electric vehicle to an existing grid-tied solar inverter. We built the prototype under a National Science Foundation Small Business Innovation Research project [xix] with subcontracts to the Larson Pennsylvania Transportation Institute at Penn State and with Customized Energy Solutions.

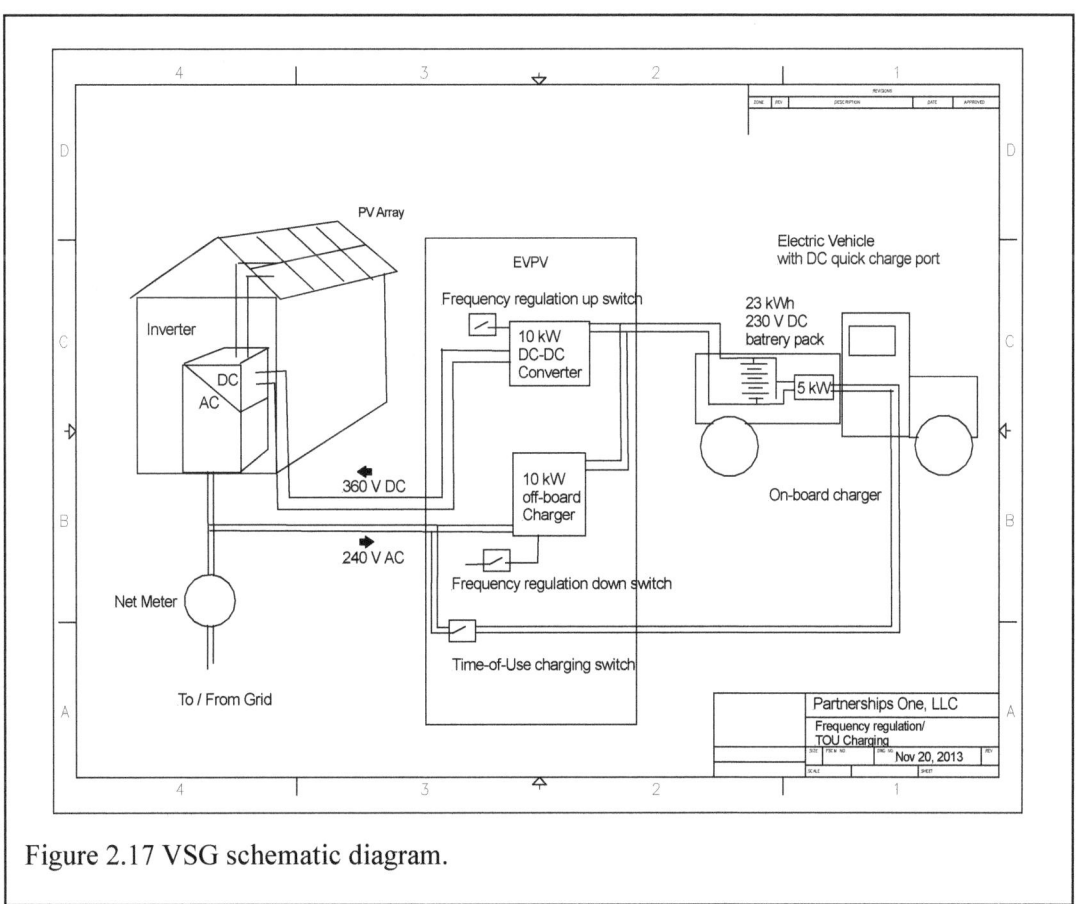

Figure 2.17 VSG schematic diagram.

There are two connections to the vehicle. One is a standard J-1772 AC connection to the 5 kW on-board charger to charge the battery. It is equipped with a Time-of-Use switch to perform charging when power is cheapest, if desired. The other connection is a quick charge DC connection to the battery pack. This could be a CHAdeMO conector as used by the LEAF and

61

other Japanese vehicles or a J-1772 level 1 or level 2 DC quick charge connector as used in some U.S. EVs. A better solution is the SAE "Combo" plug with both J-1772 and DC quick charge in a single connector now becoming standard in the U.S. and Germany. The object is to provide a high power DC path direct to the vehicle battery to enable it to take power from the grid or give it back in response to requests from the Independent System Operator.

The EV is the 2004 Ford F-150 conversion described in this manual with a 23 kWh, 230V battery and a 5 kW onboard charger. It is linked to the grid via the Electric Vehicle-Photo Voltaic (EVPV) unit containing a 10 kW off board charger for down regulation and a 10 kW DC-DC converter for up regulation, and appropriate switches and data acquisition equipment to perform regulation service and earn revenue.

The EVPV is shown in Figure 2.18 installed at the Penn State University GridSTAR facility at the Philadelphia Navy Yard with the truck plugged in to it.

Figure 2.18, EVPV installed next to net zero energy GridSTAR house at Philadelphia Navy Yard.

The VSG project is ongoing, with tests of the system at GridSTAR to demonstrate frequency regulation and the interaction between the demands of VSG and normal operation of the vehicle. To learn more about our project, visit our web site chargedupcar.com.

Frequency regulation services will saturate with an electric vehicle population of roughly 1% of all vehicles. This is an example of the massive size of the vehicle energy supply relative to the electric utilities. The next most valuable ancillary service is "spinning reserve". This is generation capacity available on a few minutes notice to shore up the utility system in the face of large increases in load or system disruptions. It is bid in the same way as instantaneous power for frequency regulation, but is not as valuable. You cannot bid for both regulation and spinning reserve with the same capacity. It is one or the other

Surprisingly, despite the enormous size of the vehicle energy demand, there is enough spare capacity in the utility system that it can supply the entire demand with existing electric generation capacity by leveling the load and charging at night. "Demand management" is a recurrent theme of utility optimization because the utilities must invest to supply the peak demand on a hot summer day, but at night the factories shut down and the lights go out. Typically night time demand is half that in the day. The utilities can supply power at night for the cost of fuel and very little else, since the investment and even the operation and maintenance is already required to satisfy the daytime peak. Thus there is a major incentive both to cut down the peak and fill in the valley. This is demand management. It has been practiced for years using night time power to heat water and for space heating with thermal storage.

Clearly electric vehicles offer a major opportunity for demand management because they are preferentially charged at night. Typically the owner doesn't care exactly when the vehicle is charged as long as it is ready to go next day. This can be accommodated by remote control of the charging process to take advantage of minimum rates at night. The technology for doing this is the "smart grid" which is being implemented to enable time of day pricing and other sophisticated demand management scenarios. As part of an optimized EV charging program this could be an important cost saving feature of V2G.

Solar Power

Solar installations at home and at work, particularly solar-shielded parking places are becoming more common. Obviously solar power can offset the increase in your electric bill when you start running an electric vehicle, (which is quite noticeable, typically 50%). If the local authorities can be persuaded to accept the vehicle as part of the routine electrical load for permitting and any rebates and Solar Renewable Energy Certificates (SREC)s that are available for additional solar power, this approach makes the most long-term sense environmentally and economically. It eliminates all the arguments about electric vehicles having a "long tailpipe" and emitting pollutants from power plants.

Current incentives for solar power make it quite attractive to power your vehicle this way over and above the desire to get off coal and nuclear power. Net metering means that the utility bills you only for the net total kWhrs that you use from them. If you send out 20 kWh during the day and take in 20 at night to charge your vehicle, your electric bill is zero. (The cost is not zero since you have to pay for the solar installation, but there are incentives there too.) This seems as though the utility is getting a bad deal because they made all the investment in transmission and distribution to get the 20 kWh to you and get nothing for it. Don't weep for them. They are getting power worth anything up to a dollar per kWh and giving it back to you at night at a cost of a few cents. Everybody wins!

Solar On the Vehicle

You can consider installing solar panels on your vehicle. Pickup trucks and vans are particularly suitable because they are big and high, and often have structures up there that can support the panels. Typically you can mount six standard 225 Watt panels in an array mounted on the pipe racks that are used to carry ladders and other long objects. The voltage of a series string of six panels will be around 300 V DC. The cost these days will be around $1250 for the bare panels and a few dollars more for some racking to support them. That's all you need because, providentially, switching power supply chargers can be operated on AC or DC. A 240 V AC charger sees around 325 V at the peak so you can wire the panels direct to the charger input with a switch to change from AC to DC. (You don't want to do both!). You will get over a kilowatt peak and over an 8 hour summer day that will take you ten miles or more. Not a lot, but not negligible.

Chapter 3. Doing the Conversion

The Recommended Conversion

Figure 3.1 Underbody of Ford F-150 pickup showing catalytic converters, exhaust crossover, heat shields and one of the sheet metal brackets under the seats which must be removed.

The first consideration after deciding to convert a vehicle using this system is to decide is which of a number of variants you wish to build. Our "preferred embodiment", to use patent talk, is a twin-exhaust modification, with a 205 V battery pack of 64 CALB lithium-ion cells, a Manzanita charger, an Evnetics Soliton Jr. controller and twin Agni 11" axial field motors. Instructions for these components will be presented as the main line of the narrative with instructions for alternatives presented at the end.

The second consideration is how much of the conversion you are willing to do yourself and how much to acquire from others. A minimum acquisition, which will save a lot of time, is to purchase the basic drive assembly and electronic enclosure and fabricate the rest from purchased or available parts. Some EV enthusiasts will have motors, controllers, etc., and more time than cash. Some may even have a complete machine shop and a desire to do it all. Others will prefer to pay for the entire job. All of these approaches are supported by this manual and the associated drawings.

Exhaust Modifications

The right side of the underbody on Ford pickups and vans (and on most others) is occupied by the exhaust system, as shown in Figure 3.1 above. This offers the best chance to make room for the electric drive system with minimum modification. All of the vehicles of interest for this conversion have either V6 or V8 engines with crossover pipes connecting both banks to a single exhaust pipe. The best modification option is to eliminate the crossover pipe (and heat shield) and redirect both exhaust streams out under the frame rails to separate mufflers under the truck bed as shown in **Drawing 1***. We have done it with one muffler (**Drawing 2**) and with two mufflers and strongly recommend the two-muffler solution, if it is acceptable to local vehicle inspection agencies. The crossover pipe constrains the space available for the drive, making for tight clearances and a difficult environment for maintenance and repair. Removing and replacing its heat shield every time you need to access the drive is a nuisance

* Dimensioned drawings, which are available in the accompanying package of **Drawings**, are shown in **Bold**.

The drawings will allow a muffler shop to perform the desired modifications to provide clearance for the electric drive system to be installed as part of this conversion. Take the appropriate drawing to your favorite shop and have them make the modifications. Trim and rebend the heat shields in the vehicle to conform to the new piping.

It is important not to interfere with the catalytic converters or oxygen sensors in the exhaust system, in part because it adds expense, but more importantly because that constitutes "tampering" with the emissions control system of the vehicle, and will void its emissions certification. That in turn will render the vehicle unable to pass emissions inspection in those states requiring it, and thus unusable and unsaleable. You may have trouble with the inspectors over the conversion in any case, and it is unwise to give them any reason to refuse it.

Sheet Metal Modifications

Figure 3.2 Single motor electric drive unit installed.

You can see in Figure 3.1 a bulbous sheet metal structure running lengthwise behind the exhaust crossover on the right side of the photo, (left side of the vehicle). You can see in the finished installation pictured in Figure 3.2 on this page that there is no room for this structure, and that it has been removed. There is another on the right side, which is obscured by the heat shield. It too needs to be removed.

These structures can be cut down to clear the hybrid drive unit, but the cleaner, simpler method is to drill out the spot welds and remove the structures entirely up to the cross member above the exhaust crossover. This can be done elegantly with a counterbore cutter. Some of the spot welds can be removed with a half inch diameter cutter, but some are larger, requiring a 5/8 inch diameter. Both can be obtained with a 1/8" pilot. Drill a 1/8" hole carefully through the exact center of the spot weld and cut the weld out with the counterbore in a ½" portable electric drill. It is also possible to drill right through the welds with a 5/8 inch drill, but this leaves a series of big holes. Once the spots are separated use a Whiz wheel or saw to cut the sheet metal projecting over the cross member joining the rear portion of the structure to the front and remove the rear portion. Some of the earlier F-series pickups, fortunately, do not have these structures.

Battery Pack Safety Considerations

WARNING---EXTREMELY--- SERIOUS---HAZARD

The batteries are grouped in modules to reduce the risk of electrocution and arcing, which is otherwise very real with these high voltage DC packs. DC electricity, Unlike AC, does not have current zeros which allow short circuits to clear. Once a DC arc is struck it will keep going. The stored energy in the battery pack amounts to **72 mega-joules,** equivalent to **sixteen pounds of TNT**, and it can be released almost as suddenly.

The National Electrical code favors DC systems at less than 50 V, which are safe. One hundred Volts of DC will give you a very nasty shock. **More than 100 Volts is very likely to kill you. Treat it with great respect. You will only make a mistake once.** Test every potential hot spot with a meter before you touch it or put a tool on it. Tools should have insulated handles or be wrapped in electrical tape. **Never** put tools or other hardware down on a battery bank or where they can fall or roll onto the terminals.

Battery Installation

We have used Group 24 and Group 27 AGM lead-acid batteries and Trojan 1275 flooded lead-acid batteries to power the prototype trucks, as well as Group 24 Valence lithium-ion batteries and 100 Ah individual prismatic lithium-ion cells grouped in 25 V strings of eight cells. The batteries were grouped in nominally 48 V modules, connected by fuses and contactors as shown in Figure 3.3 below. The reason for this is to make the individual modules easier to handle than the entire pack, which may weigh half a ton, and more importantly to separate the pack into nonlethal modules to eliminate the shock hazard until the truck is actually in operation, and the high voltage wiring is safely contained in an interlocked compartment. The truck bed is 62 inches wide at the top, 65 inches at the bottom, and the long bed version has 28 inches of full width at the bottom before running into the bulges for the gas tank filler and the rear wheel wells. Four 48 V modules fit into this space as shown in **Drawings 3 and 4**.

Most of our recent experience is with packs of individual lithium-ion cells which offer flexibility and lower cost at the expense of more complex and expensive installation, questionable life, poor quality control and poor product support. We recommend consideration of Valence Technology group 27 batteries which will store up to 28.3 kWh at 205 V for ease of installation,

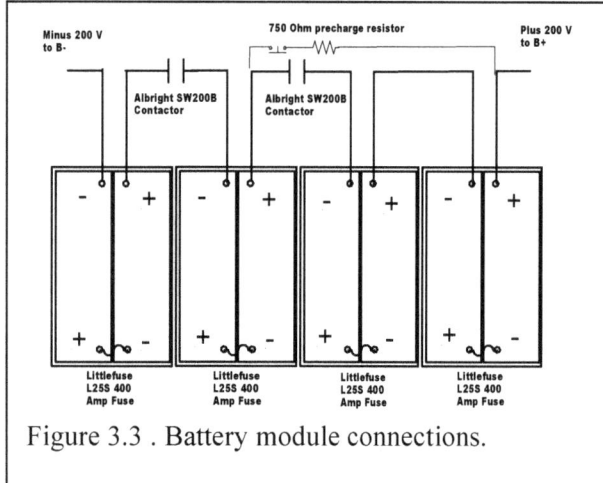

Figure 3.3 . Battery module connections.

reliability, and good product support. As mentioned in Chapter 1, we have had good experience with Valence batteries in demanding marine service. The one problem has been that their Battery Management System is so protective of the batteries that it is sometimes a nuisance. This is compensated in a PHEV in that you can always run on gasoline till you straighten out the electrical problem. Both Valence group 24 and group 27 batteries have a maximum continuous current limit of 150 Amps. This is what you will draw at a steady 60+ mph on the interstate. At lower speeds you will draw up to 200 A but for shorter times. This is somewhat marginal but probably OK. The Valence battery installation details are given below as an option

Individual Lithium-Ion Cell Installation

Battery mounts are fabricated from angle iron to secure the batteries firmly in the vehicle. In a crash several hundred pounds of batteries on the loose could be very dangerous, not to mention the electrical fireworks that can be released with explosive violence. The Tour de Sol competition had a requirement that batteries be mounted to withstand a 20 G acceleration forward or backward and 10 G vertically or sidewise. This is a desirable standard. Fortunately, pickup truck beds are already designed to withstand heavy loads in a crash, so the objective is to fasten the batteries firmly to the bed. The batteries should be mounted in the extreme lower front end of the bed where it is particularly strong and protected in crashes.

The increasingly common and inexpensive individual lithium-ion cells from Chinese manufacturers such as Thunder Sky and CALB are becoming standard for vehicle conversions. Sixteen cell modules of 100, 130 and 180 Ah CALB lithium-ion batteries in their mounts are shown in **Drawing 3**. These will fit the truck bed with capacities of 20, 26 and 36 kWh at 205 V. Four twenty-cell modules of 130 Ah, cells will give 33.3 kWh at a maximum of 252V.

The individual CALB cell dimensions are:

	Width	Height	Thickness
100 Ah	142mm(5.59")	218mm(8.58")	67mm(2.64")
130 Ah	182mm(7.17")	278mm(10.94")	56mm(2.20")
180 Ah	182mm(7.17")	278mm(10.94")	71mm(2.80")

An 800 lb 180 Ah battery pack at 20 G will impose a shear force on the fastenings of 16,000 lb. Because it is almost impossible to access the underside of the bed over the gas tank, self tapping 5/16" steel cap screws can be used there. Elsewhere 5/16" grade 5 cap screws with nuts and fender washers under the bed should be used. Six screws per module are sufficient. The screws should be located at the ribs on the bed where the battery mounts are in direct contact with the sheet metal. There are twelve ribs on 5" centers with a total width of 65 inches. Each battery module should bridge and be fastened to three ribs with space between the modules for clearance and cooling as shown in **Drawing 3**. If a bed liner is present, it should be removed under the battery boxes to get metal-to-metal contact for mounting. Washers should be used under the screw heads.

The batteries themselves must be securely fastened to the battery mounts. A minimum of two 3/8" (10 mm) screws per eight cell battery provide a safety margin of 100%. **Drawing 3** shows a frame around the top of each stack of eight cells with fiberglass angle sides for electrical isolation and angle iron ends to locate the 3/8" x 9" hold down screws which run from top to bottom, threaded into nuts welded to the battery mounts. Installation of the batteries themselves should be postponed until all the drilling and fastening are finished to avoid accidents.

The individual lithium-ion cells need a battery management system to measure the voltage of each cell and preferably the temperature of each cell also, so that the cells can be equalized during charging and so that failures can be detected. As mentioned in Chapter 2, lithium-ion cells must not be overcharged or over discharged to avoid irreversible damage to these very expensive batteries. The battery management systems available are essentially bare boards which cannot be exposed to the environment and must be mounted in the electronic enclosure. This makes for a mass of wiring if the cells are mounted outside the enclosure as is required for lead-acid batteries and acceptable for Valence batteries. A preferred mounting arrangement for individual lithium ion cells is to enclose everything, the cells, the BMS and the power electronics, in a specially made aluminum box shown in **Drawing 5**. The box should be drilled for all the fastenings and installed after the battery mounts with four sheet metal screws into the truck bed.

Battery Pack Wiring and Installation

The power cables to the right side motor as well as the 12V control circuitry and the instrument lines are all routed through a 2" conduit through the right side of the truck bed. Liquid-tight conduit (McMaster Carr #8071K47) is more rugged, Ultra-flex (McMaster 8069K17) is easier to install. A 2 3/8" diameter hole is hole-sawed in the right side of the truck bed 8 1/4"" back from the seam at the front of the truck bed and 11 to 16" up from the bottom where the side is flat. The conduit goes through this hole by means of straight conduit fittings

(#7119K77) in the bed, a 2" pipe coupling and a 90 degree 2" elbow (#7119K96) inside the fender. For lead-acid and Valence batteries the hole is accessible in front of the cross bed tool box containing the power electronics. For individual lithium-ion cells a matching hole in the enclosure gives access.

The conduit from the outside elbow to the right side motor is 42 inches long for the single motor installation and 56 inches for the twin motor installation. It should be installed next by reaching up inside the fender to tighten it onto the outside elbow. The conduit ends in a 2"elbow that feeds into the top of the right hand motor box to bring all the cabling into the box with a weather tight seal. Wiring to the motors is described below.

The individual cells now need to be mounted and connected together in series. This connection needs to be very secure because the full motor current is passing through each of the connections, and any loose connection will get hot and fail, frequently to the extent of melting the battery terminal involved. The wire (cable or buss bar) itself needs to be adequate in size to carry the current and flexible enough to withstand installation and vibration without breaking. The insulation needs to be able to withstand the environment which frequently involves chafing, water and exposure to oil and gasoline. Stranded cable is mandatory, and the best is welding cable, which is fine-stranded, high quality cable with ample, high quality insulation.

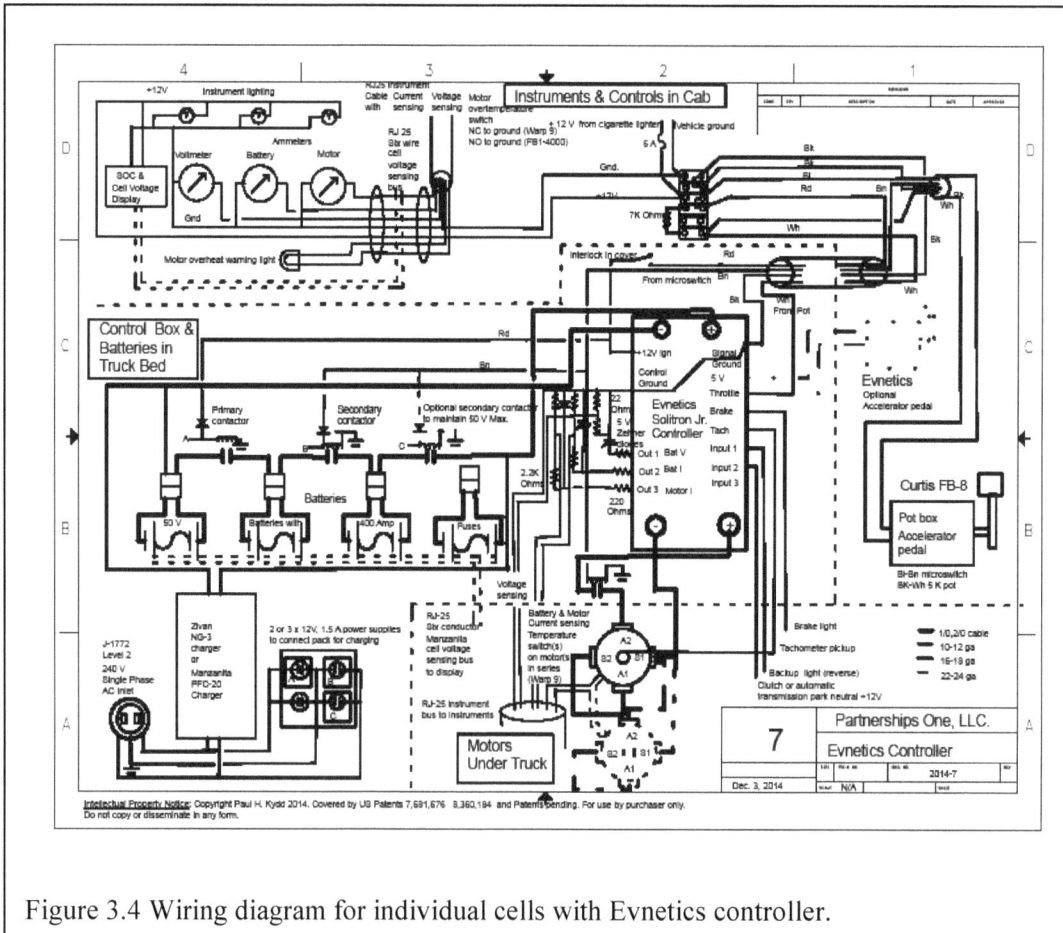

Figure 3.4 Wiring diagram for individual cells with Evnetics controller.

The battery and motor current drain are limited to 300 Amps by the controller, which is not particularly high in electric vehicle terms. The total length of high current series circuit is approximately 52 feet, comprising 28 ft in the battery connections shown above plus 6 feet in the control enclosure and 18 feet to the motors under the truck. The resistance of #1 cable is 0.127 ohms per 1000 feet which will give a 1.32 V drop at 200 amps and waste 264 Watts out of 40,000 delivered (0.66%). 1/0 cable has a resistance of 0.099Ω per 1000 ft and will give a 1.03 v drop. 2/0 cable has a resistance of 0.077 ohms and will cut the loss to 0.8V at the cost and weight penalty of twice as much copper as #1. The individual cells are often connected with copper buss bar which should be at least 20 mm x 3 mm (3/4" x 1/8"). 1/0 cable is a good choice for the rest of the battery current circuit, while 2/0 is the Cadillac treatment for the motor wiring. The wiring diagram is shown in Figure 3.4 above.

The most satisfactory connection for heavy current wiring is the crimp lug. This is a massive copper termination with a flat section drilled to match the screw fastening on the battery or other component and a tubular section that is a close fit to the wire size chosen. Quality lugs are heavy-walled, accurately matched to the wire diameter, and plated to minimize corrosion. It is almost mandatory to cut the cable with a shear having semicircular jaws that cut cleanly and compress the strands into a circular bundle. Otherwise it is very difficult to get the cable into the close fitting lug. To make a crimp connection, strip the wire for slightly longer than the length of the tubular section, insert the wire, and crimp the tube down on the wire firmly. On the larger wire sizes it is best to crimp near the open end of the lug and crimp again toward the closed end to compress the wire end trapped by the first crimp. The best results will be obtained with a long handled crimping tool with rotating anvils to fit various size lugs. A cheaper solution is to use a hammer crimping tool which has a hardened punch which you drive down on a V shaped anvil with a two pound hammer. A proper crimp will compress the wire strands and even partially cold weld the strands together and to the lug. It will withstand mechanical abuse and overheating. Soldered joints are not recommended because they will fail when heated.

Large semiconductor fuses such as the Littlefuse L25S-400 Amp are used to provide circuit protection. A single fuse in the series circuit is mandatory. More than one is overkill from a system point of view, but is recommended in case somehow a single module is shorted by a dropped tool or otherwise, either in or out of the vehicle. The fuses with suitably formed copper buss bar segments make convenient connectors between the 25 Volt series strings to complete the 50 Volt modules, as shown in Figures 3.3 and 3.4.

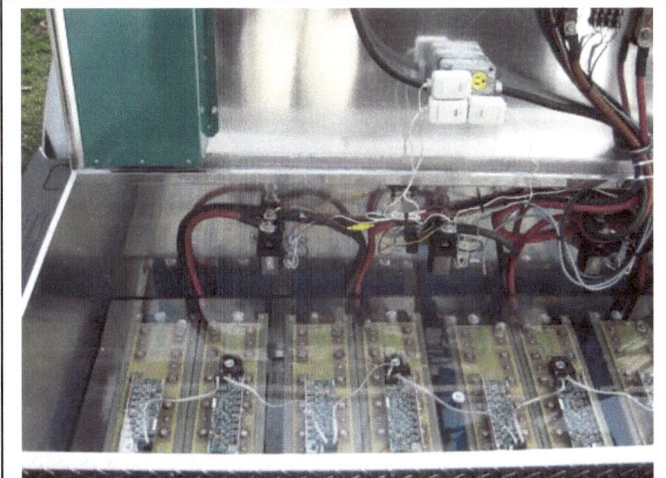

Figure 3.5   Interior and lid of lithium-ion battery enclosure

Power Electronics Installation

The enclosure for single cell lithium-ion batteries and related power electronics is a fairly major item. If you have access to a good sheet metal shop you would be well advised to pay them to fabricate it from **Drawing 5**. The holes for attaching the electronic parts to the lid are not so difficult and you can do that yourself with an electric drill per the dimensions in the drawing. The Evnetics Soliton Jr. controller and Manzanita micro charger are recommended.

The three or four contactors mount on the front (hinged) side of the box as shown in Figure 3.5. It is convenient to use ¼-20 rivet nuts for this because it is impossible to assemble them single handed, and if you ever have to take them off, conventional nuts are inaccessible. All of the battery connections and the BMS connections are on the cells in their battery mounts which are installed in the truck bed first. The enclosure with holes drilled and rivet nuts installed is slid over the battery mounts and fastened to the truck bed with sheet metal screws. The cells are then installed and wired direct to the Albright contactors without the need for Anderson connectors. For safety, however, make these connections to the contactors last, so that all of the power electronic wiring is done before the system is hot.

CALB lithium cells can be connected together through a specially manufactured printed circuit that saves a substantial amount of labor in connecting the individual cells and the voltage and temperature sensors that go to the battery management system. Manzanita Micro of Kingston, WA, supply regulators which are battery management systems for 4 and 8 cells. They provide equalization during charging and can be interfaced to a computer or their own display unit to display the voltage of individual cells. The regulators plug in to "Reg decks" which are large printed circuit boards providing all of the electrical connections needed for voltage monitoring of a series string of 8 cells. They also provide thermistors, which contact the cells to monitor their temperature. Finally, they have massive copper buss bar segments to connect the series string of cells into a battery. These decks avoid a huge amount of hand wiring and cabling and provide a neat solution to cell aggregation. They and the regulators must be protected against the weather as the connections are all in the open and connected electrically. Water on the boards will lead to instant corrosion of anything connected to the positive poles of the cells.

The regulators are chained together with RJ-25 six-conductor telephone cable. You will need to get a crimper and a package of 6 connector terminals to make short jumpers out of a single length of RJ-25 cable. One end of the chain goes to the charger so that it can reduce voltage and current if the cells become over charged. The other end goes to the display on the dashboard to show the driver the voltage and temperature of each cell, and to allow programming of the regulators to set maximum and minimum voltages, etc

The batteries are covered by a sheet of 3/16" thick Plexiglas shown in **Drawing 5** to provide additional protection against condensation dripping from the lid of the box while allowing one to monitor the LEDs on the BMS units. The plexiglas also provides thermal insulation to keep the batteries warm in the winter. It cannot fit too tightly because cooling air for the charger and controller come up from the gaps between the bottom of the box and the grooves in the truck bed. Approximately 12 square inches of opening is needed. Finally the Plexiglas provides electrical insulation to prevent casual viewers from contacting the battery terminals or dropping tools on them. A periodic survey of the high current connectors with an infrared scanner is highly advisable immediately after running the truck to detect loose connections. Another Plexiglas panel should be installed on standoffs on the cover to prevent shock hazard from exposed connections on the controller. Be sure not to restrict access to cooling air for the Evnetics controller.

DC Contactors

Aggregation into a single high voltage pack is accomplished by connecting the individual modules with a minimum of two Albright contactors as shown in Figure 3.3. At least one primary contactor is absolutely required to allow for shutting off the battery pack remotely. The relay for the precharge resistor recommended for Curtis controllers is wired in parallel with the primary contactor so that when the system is turned on the big capacitor bank in the controller is charged up over a few seconds rather than drawing a huge short circuit current through the batteries and the secondary contactor on closure. The precharge feature is built in to Evnetics controllers and an external resistor is not needed.

When the primary contactor is closed the entire system is at full voltage and should be protected by an interlocked mechanical cover which will cut off the primary if the cover is opened. Even after cut off it will take many seconds for the capacitors to discharge, so treat it as hot until proven otherwise. In addition to the electrically-operated contactors, a manual cut-off is recommended. You can buy an Anderson 350 Amp connector pair in a sheet metal bracket with a handle to allow for emergency manual disconnect with one hand or a lanyard.

The secondary contactor between modules 2 and 3 is actuated by a microswitch on the accelerator pot box to actually start the flow of electricity. Yet a third Allbright contactor actuated in parallel with the secondary is required between the controller and the motors. When the truck is being driven in reverse by the IC engine, the motors generate a current which is short circuited by the freewheeling diodes in the controller causing a drag on the vehicle and a lot of stress on the electric drive. The third contactor prevents this. Permanent magnet motors such as the AGNI generate high voltage when being driven in either direction, and the motor contactor is especially critical for them.

The contactors should be large and able to interrupt high voltages and high DC currents. Allbright contactors are often used in the 200 Amp size with magnetic blowout capability to help extinguish the arc. It is important to install these contactors with the right polarity for the blowout function to work properly. It is highly recommended that you use the protected "P" designated contactors, which are enclosed to keep dirt, debris and bugs (literally) out of the contacts and ensure higher reliability.

On Board Battery Charger

The charger and the controller are now mounted on the lid of the box shown in **Drawing 5**. Their weight helps hold the lid down, and the direct connection provides for heat dissipation through the lid. The charger and controller need exhaust vents in the rear side of the box in addition. The cooling air vents should be protected by two 4 " diameter louvered and screened covers on each side to allow warmed air to exit while preventing rain and bugs from entering. Orienting the charger toward the rear as shown ensures that the indicator lights and controls on the Manzanita charger are accessible from inside the truck bed when the lid is opened. Be sure to waterproof all fastenings in the lid with RTV silicone rubber to prevent rain water from entering the enclosure.

The recommended charger is the PFC-20,-30 or-40 series from Manzanita Micro 26125 Calvary Lane NE, Suite 300, Kingston, WA 98346. They offer versions at 20, 30 and 40 Amps maximum AC current which is adjustable by a front panel potentiometer. The output voltage is adjustable from 12-450VDC by a 20 turn pot. Newer models offer digital voltage adjustment, which is highly recommended. The time at constant voltage is also adjustable. This is all very convenient if you have to charge part of the pack to equalize it, or perform some other modification without having to send the charger out to reprogram it. We have used the PFC-20 model to charge a 20 kWh pack very satisfactorily, It gives a full charge in 6 hours. Another flexible feature of the Manzanita charger is that it will operate at either level 2, 240 VAC or level 1, 120 VAC with no adjustment, although the AC current limit means that charging at 120V takes twice as long.

The charger should be wired as shown in the wiring diagram, **Drawing 7.** It is connected direct to the B+ and B- poles of the battery at the controller. The current can be as high as 40 amps so #12 wire or heavier is indicated. Under no circumstances should this connection be fused. The Manzanita demands a battery connection, and if it is disconnected, the charger will destroy itself through over voltage. It is internally fused for circuit protection. If used with Valence batteries and the Valence BMS which cuts off the charger if the batteries are fully charged, the cut off contactor will have to be in the AC line to the Manzanita charger, not in the DC output.

For charging, the primary and secondary contactors on the battery bank must be closed to complete the series string. The charging voltage can be provided by plug-in twelve volt, 1500 mA Radio Shack power supplies powered from the 240 V input by taking 120 V from one line or the other and the grounded neutral as shown in **Drawing 7, and 7A.** The neutral should be grounded to the box. The 12 V running power should be supplied through diodes on each contactor coil to isolate the charging control signal from the running control signal to avoid overloading the charging power supplies which can handle one big contactor each.

The 240 V input should be through a male receptacle mounted on the rear (tail gate side) of the box. Preferably this should be the SAE J-1722 standard inlet for electric vehicles, which is now available from a number of vendors, along with the matching female plug reminiscent of a gasoline fuelling nozzle. If these options are too pricey, a watertight NEMA L6-30 turn lock receptacle and plug pair can be used. In either case it is more convenient to have the cable attached at the house end and a flush receptacle on the truck, rather than having the cable on the truck. A separate cable assembly can be carried to permit charging from the popular NEMA 10-50 (clothes dryer), NEMA 14-50 (four-wire grounded) or NEMA 6-50 (welder) receptacles.

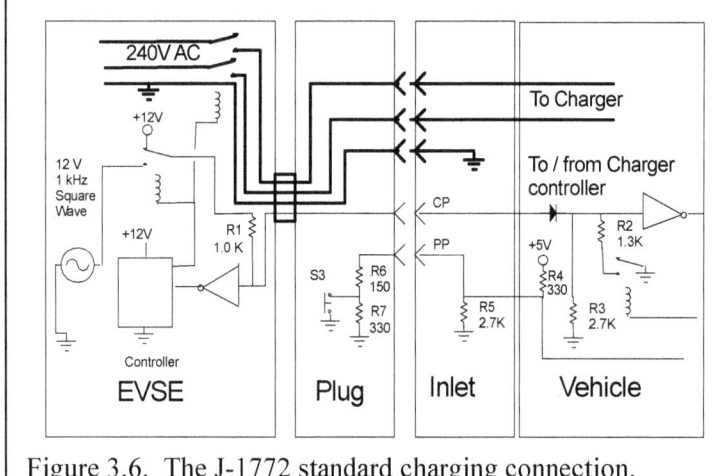

Figure 3.6. The J-1772 standard charging connection.

SAE J-1772 Inlet

The J-1772 standard provides for exchange of information between the vehicle and the Electric Vehicle Supply Equipment (EVSE). There are five pins in the plug and receptacle as shown in Figure 3.6. Two provide 240 V (level 2) AC power and one is a ground. The other two are light duty signal connections designated Contact Pilot (CP) and Proximity Pilot or Plug Present (PP). The vehicle senses the plug being connected via a voltage drop from 12 V to 9V due to current flow through the diode to the 2.7K resistor R3 and turns on the square wave generator. The vehicle then adds 1300 ohms dropping the signal voltage to 6V to start charging. The vehicle control system can add a parallel 270 ohms to drop the signal to 3V if the vehicle needs ventilation for lead-acid batteries and can vary the resistor to indicate to the EVSE that it is capable of higher charge rates. Switch S3 is connected to the latch on the plug which adds 330 ohm R7 to the 5 V connection to signal to the vehicle to shut down the charger if the plug is pulled. The PP connection also should deactivate the controller so that the vehicle cannot be driven away while plugged in. To make your conversion minimally compatible with the standard you will need to connect the CP pin to ground with a diode and an 880 ohm resistor substituting for the parallel 2.7 K and 1.3K resistors of the full system.

The Controller

The Evnetics Soliton Jr. controller shown installed in Figure 3.7 has a capacity of 340 V, 600A peak and 450A continuous. It can be liquid cooled for high currents, but has fans allowing operation at lower currents, air-cooled. It does not need a heat sink. It can be programmed through an Ethernet cable to a laptop with a web page format. The following parameters can be set:

Minimum battery voltage at no current and at full current
Maximum battery current
Maximum motor voltage, current and power
Slew (ramp) rate

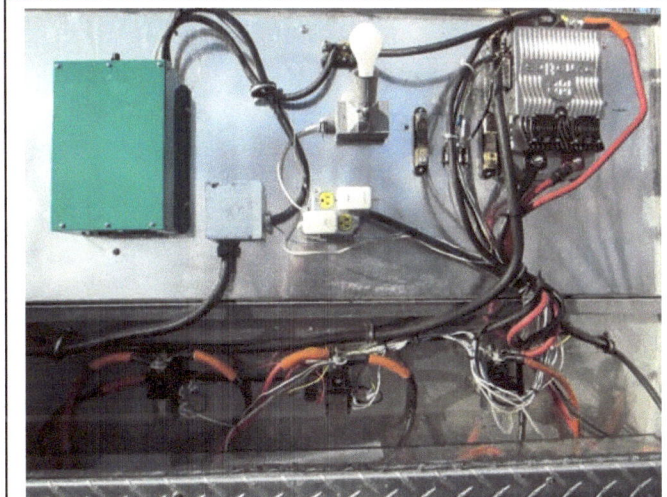

Figure 3.7 control installation with Evnetics Soliton Jr, controller and Manzanita PFC-20 charger. 12 V supplies to connect the pack during charging and a 150 W light bulb for maintaining capacity during the winter also shown.

There are inputs for throttle, brake and tachometer. The throttle input requires a 0-+5V signal. There is a +5V output and the Evnetics throttle unit is a potentiometer which uses this voltage to provide the required 0-5 V throttle input. It is also possible to use the Curtis or other 0-5K pot box by taking +12 V through a 7K resistor to the throttle input and through the Curtis pot to ground as shown in **Drawing 7**. When the Curtis throttle is at 0 ohms the throttle voltage is also 0V. As the pot comes up to 5K the voltage rises to +5V.

There is a choice of PWM frequencies, 8 kHz for performance or 15 kHz for quiet operation. The brake input inhibits motor current when connected to the +12V vehicle brake light circuit. You may feel that the combination of brake and power is so unlikely that this connection is not warranted.

The tachometer pickup is used primarily to set an idle speed for full electric conversions to operate the pump in an automatic transmission and an overspeed limit for the same, which is not really relevant for supplemental PHEV conversions. The tachometer can be programmed for magnetic pickups providing one to six pulses per revolution on the motor to set a maximum motor speed limit to prevent overspeed damage to the motor(s).

The Evnetics also has three digital inputs. Each of the three inputs can be set to off, throttle limit, reverse or start. When applied to input 1 and configured in "reverse" the back up light signal will prevent electric motor operation in reverse. Input from the transmission park-neutral start switch on the vehicle to input 2 configured for "throttle limit" and with the throttle range set to 0% on the web page will prevent electric motor operation in neutral and park while starting the IC engine. Using this option will require an override manual switch to open the circuit and allow electric operation in neutral for automatic transmissions that do not have free wheeling. Input 3 could be used in conjunction with the "Run" output to enable throttle operation, or in conjunction with an over temperature switch on the motors to shut off power and let them cool down while windmilling.

There are also three outputs which can be configured as meter drivers for motor current, battery current, power, controller temperature or battery voltage. The meter outputs are 0-5 V and require 100:1 voltage dividers to drive standard 50 mV full scale meters as shown in **Drawing 7.** An output configured as "tachometer" can convert whatever ppt signal comes from your sensor to the equivalent signal to drive a conventional tachometer, e.g. 1ppt from the motor to 3 ppt to drive a six cylinder tachometer. An output configured as "water pump" can be used to turn on liquid cooling at a controller temperature of 40°C and turn it off at 35°C. Pretty nifty!

Control Enclosure-to-Motor Wiring

The cables to the motor(s), the control wiring and the instrument wiring should project the proper distance to pass through the 2" conduit to make connections to the motor(s) and to the receptacles for the control and instrument wiring in the motor electrical box. The 2" conduit ends in a 2"elbow that connects to the box on the back of the right side Agni motor, shown in Figure 3.8.

The control and instrument wiring is fed down through the 2" conduit conveniently as two four-conductor 18 Ga. cables The first of these is for brake light, tachometer, reverse, and clutch/transmission signals, i.e. those signals coming from the vehicle to the controller. This cable will terminate at a four-pin female receptacle (McMaster Carr 3214K29) on the back of the right hand motor box. Typically these receptacles have a lock nut on the back, and to make them removable without cutting all the wires, a bracket is needed that is screwed to the inside back wall of the motor box allowing the receptacle to project through the hole for connection. Unscrewing the sheet metal screw that holds the bracket allows the receptacle to be pulled back into the motor box and back through the 2" conduit for motor removal. About twelve inches extra of four-conductor cable will be required to allow the receptacle to reach the hole comfortably,. The connection from the motor box to the transmission will be made with the corresponding male plug and cable (McMaster Carr 3214K27). About five feet of cable will be needed to reach the transmission. Up to seven feet of additional wire is available to connect to the brake light circuit and the tachometer pickup.

The other four-conductor cable is for black, white, red, and green wires coming from the controller in the truck bed to the 12 V supply and the potbox in the cab. It too should be terminated by the same type of female receptacle in the same sheet metal bracket. The control and instrument wiring will be continued to the cab by the use of the same type of plug with a12 ft. cable. The cable will feed up through the existing hole in the floor of the truck just behind the passenger's seat and extend forward under the seat to reach the junction box mounted on the transmission hump.

The instrument outputs require six conductors which are most conveniently and economically purchased as flat six-conductor RJ-25 telephone-type cable These can be purchased economically as 12 ft runs with plugs on each end. If you use the Manzanita display you will need another RJ-25 six-conductor cable for that. Both of these cables come down through the 2"conduit from the control enclosure and need to terminate in receptacles for similar cables to continue on to the cab. The neat way to do this is to purchase pairs of receptacles such as Mouser part no 592RJ116N3B and mount them on a small piece of circuit board supported inside the right hand motor box on standoffs. Be sure to get the wiring right so that blue connects to blue, etc. The cable ends coming down from the control enclosure are plugged in to the circuit boards, and the continuing cables leading up to the cab are fed in through the grommeted drain hole in the bottom of the motor box and plugged into the paired receptacles. In this way the entire assembly can be disconnected for motor removal, while the relatively fragile RJ connectors are inside the box protected from the under-truck environment. The receptacles mentioned have LC filters built in which will help to eliminate RF noise from the controller. The continuing cables will be fed up through the same hole in the floor of the cab as the four-conductor control cable.

The Poly Chain Drive

Figure 3.8 Agni twin axial flux motor drive ready for installation

The twin motor drive with Agni axial flux motors, is the next major component to be installed. The drive itself comprises the motors mounted on a support structure which connect them to a Poly Chain drive unit, as shown in Figure 1.16 from the front and Figure 3.8 above from the rear. The drive is shown in relation to the truck in **Drawings 8, 9, 10** and in detail with dimensions in **Drawings 11 and 12,** The drive unit applies the torque from the motors to the vehicle's drive shaft in a manner that leaves the latter free to vibrate and move about as it did before conversion. The result is a drive that applies a pure torque to the drive shaft sprockets but doesn't restrain them from moving with the vibrations and rocking of the IC engine and transmission. This arrangement reduces stress on the rear bearing of the transmission.

Figure 3.8 shows the drive unit ready for installation. The two Agni 11"motors are mounted on a formed 3/16" thick steel cross bar with 3/16" steel brackets. The brackets also support the twin 32 tooth idler assemblies which line up with the 53 tooth Poly Chain sprockets on the motors. The 67 tooth sprocket mounted on the drive shaft was shown in figure 1.17 and the mounted drive in Figure 1.18.

Assembly of the drive begins with assembly of the motors to the support structure, which is designed to be installed and removed as a complete-prealigned unit. The support structure has welded studs which support the idler assemblies and holes for mounting the motors and the idler support posts as well as the tensioning cable pulleys. Use six M10 x 20 mm grade 5 cap screws and lock washers to assemble each motor to its support bracket.

Next the idler assemblies are installed. The idler sprockets are assembled to the idler brackets with specially modified hollow screws. The screw heads are facing away from the motors and the asymmetric idler sprockets are installed with the sprocket side close to the motors and spaced with ¾" SAE washers on either side to allow the bearings to turn freely. The idlers are secured with 11 mm thick ¾-16 jam nuts and internally toothed lockwashers. The nuts for the lower assemblies are notched for tensioning cable attachment. The upper idlers have a lever extending away from the idler sprocket itself and the lower idlers do not. The tensioning cables connect the pair of idlers on one side to the pair on the other with a turnbuckle to tighten the chains. The cables pass over 1 3/4" diameter ball bearing pulleys mounted on ½" x 3/8-16 shoulder screws as shown in **Drawing 8.**

The lower idler units are supplied with balancing springs which should be installed on the studs before the units are mounted. Slide the idler assemblies onto the mounting posts with some lithium grease to lubricate them. The springs are calibrated to counterbalance the weight of the lower idlers, and, through the cable assembly, the upper idlers as well. Wind the balance springs up one turn and insert the hooked ends into the holes in the motor mounting brackets. Ensure that the lower and upper units move freely around the studs. The springs will be snug around the base of the idler mounting studs when properly installed.

Install the idler support brackets on each pair of posts with 12 mm x 100 mm cap screws, nuts and lock washers. These supports lock the pairs of idlers together and make a much stiffer structure to resist the very large forces imposed by the drive. Finally the idler posts are secured to the top of the support bracket with ¾-16 nylon-insert, low-profile lock nuts.

Next the tensioning cables should be installed. Tie the idlers to the post of the triangular support structure to hold them in their minimum extended position. Each 3/32" stainless steel cable engages the notched nut on its respective lower idler assembly by a tight loop made with a Nicopress sleeve. A Dremel tool with a cutoff wheel makes a nice clean cut in the wire so that it can pass through the sleeve. Pull these loops as tight as possible before compressing the sleeves. A special compressing tool makes this easy, but is quite expensive. A simple bolt-together tool can be used. The two cables from left and right lower idlers then pass around the 1 3/4" pulleys and through the ends of the turnbuckle let out to its maximum extension with just four turns of thread engaged at each end. A 3/32" thimble and a Nicopress sleeve should be slid onto each cable before passing around the turnbuckle pin. The cables pass back through the sleeves but are not tightened yet. The cables continue back around the 1 3/4" pulleys and up to the holes in the upper idler levers where they are each secured with another thimble and Nicopress sleeve, which is compressed under tension to minimize slack in the cable..

When both cable loops have been competed, tighten up the turnbuckle against the ties to the posts to get the cable runs even and compress the Nicopress fittings at each end of the turnbuckle to keep them that way. If done properly, the idlers should move out evenly from the center when the turnbuckle is tightened to tension the Poly Chains, while allowing the 67 tooth driven sprocket to move from side to side as the turnbuckle moves freely right and left with the idler mass supported by the balancing springs.

Next install the plastic electrical box on the right hand motor to accept the electrical connections and protect them from the environment and the laminate insulating plate on the left hand motor.

Installation of the Drive

The first step in installation is to remove the drive shaft from the truck and install the 67 tooth Poly Chain drive sprocket as shown in Figure 3.9. A lift allowing you to work under the truck standing up is almost essential for this step, and well worth the trouble and expense to rent or borrow. Alternatively have a repair shop perform this step for you.   To pull the shaft, loosen

Figure 3.9 The drive shaft installed with sprockets and twin exhaust modifications ready for installation of the drive

the 12 point screws that secure the rear universal joint to the differential. There is a special flexible jointed 12 mm socket wrench that makes this job easier. Mark the shaft and rear axle flanges so that they will go back together the same way. The truck has a two part shaft with a center bearing. Loosen this now with an impact wrench, separate the rear flange and carefully pull the shaft back out of the transmission. You will see a splined shaft sticking out of the transmission. It is desirable to have a 1.60" diameter hollow plug ready to fit over this shaft and seal the back of the transmission to prevent transmission fluid from leaking out.

Once the drive shaft is on the bench, assemble the 67 tooth by 62 mm sprocket and its distance piece to it. The sprocket should be bolted to the distance piece with six 5/16-24 grade 8 high strength screws alternating with 5/16" roll pins as shown in **Drawing 12**. Slide the sprocket assembly onto the internally splined forward yoke of the front universal joint until it seats firmly against the round boss of the vibration damper ring on the yoke. Be sure that the distance piece does not hang up on the radius on the yoke. Clamp the sprocket assembly to the damper ring so that it cannot move in the subsequent drilling operations and check that it is absolutely perpendicular to the yoke. Drill the damper ring in six places with a #4 center drill. The two opposite places where the yoke is thin are drilled 19/64" and reamed to fit 5/16" dowel pins for alignment. The other four holes are drilled with a letter I or 17/64" tap drill and tapped 5/16-24 for four 5/16-24 x 1" grade 8 cap screws and lock washers.

The sprocket assembly should be made with care and should come out balanced, but for the best quality job, have the shaft balanced with the sprocket in place to ensure against vibration. The distance piece and sprocket assembly can be checked for runout at the same time. Drive shafts turn fast under heavy load and should be installed with care.

Reinstall the drive shaft with the two 1120 mm Poly Chains looped over it. Carefully work the splined joint on the front yoke onto the transmission shaft. The rear flange should come together just as it came apart, and if neither the engine nor the rear wheels have been rotated, it should be just as it was originally. Refasten the center bearing with adequate but not excessive torque as the bolts are not the strongest and refasten the rear flange with the twelve point socket to the recommended torque.

Using a transmission jack, raise the drive unit with the motors until the cross member is in light contact with the bottom of the frame rails and adjust its position until the flanges on the idlers line up with the 67 tooth sprocket, i.e. the front flange on the front idler lines up with the front flange on the driven sprocket and the rear flange on the rear idler with the rear flange. Install the mounting clips over the frame rails and fasten with 3/8-24 grade 5 screws. It is important to follow the procedure outlined and under no circumstances drill holes in the frame, which is an engineered, high-strength member.

Work the 1120 mm Poly Chains forward and install them on the 67 tooth sprocket and then stretch them over the idler assemblies. Lastly, the unflanged 53 tooth drive sprockets are installed. Push the sprockets and their taper lock bushings and keys onto the motor shafts while aligning the teeth with the chains. When the driving sprockets are aligned axially with the idlers, tighten the screws in the taper lock bushings to secure them. It is desirable to use Locktite on the shaft key seats to make sure that the keys stay in.

Tighten the turnbuckle and cable assembly to tension the Poly Chains. They should be tight enough that the chain deflects approximately one chain depth (6mm) under a ten pound perpendicular load, i.e. pretty tight.

Pull the electrical cabling through the 2" elbow on the right hand motor box and secure the elbow with its locking nut. Connect the power cables to the motors in series. Mount the receptacles for the control wires through the ½" NPT holes in the lower rear of the box with the bracket and sheet metal screw. Connect the red, green, white and black wires to one receptacle. Connect the wires leading to the brake, tach, reverse, and transmission on the controller to the other receptacle. Plug in the two four-conductor cables and run one forward to connect to the transmission. The reverse switch output wire is dark green and yellow. Connect to it with a crimp-on fitting. The transmission park-neutral start wire is tan/red. Connect to it with a crimp-on fitting. Plug the instrument and monitor RJ cables into their respective receptacles on the mounted circuit board in the motor box. Connect the overtemperature wires if supplied on the motors to the appropriate jumpers on the circuit board. Run the RJ cables going to the cab through the grommeted drain hole in the bottom of the box and plug into the mating receptacles. Find the hole in the cab floor behind the right hand seat and make a 3/8" hole in the floor mat centered on the hole. Run the instrument and monitoring cables up through the hole and forward under the seat to the junction box on the transmission hump. Last, plug the four conductor cable running to the cab to the respective receptacle and run the cut end up through the hole in the floor mat and forward under the seat. Install the cover on the box. This completes the work under the truck.

Inside the Cab

The major components inside the cab are the accelerator pedal-pot box, the junction box on the transmission hump to connect all the wiring, and the instrument package..

The wiring diagram **Drawing 7** (Figure 3.4) shows a Curtis FP-8 accelerator pedal assembly which contains a 5 kΩ potentiometer and a microswitch. The Evnetics potbox is similar in function but without the microswitch and the pedal. Despite Evnetics' disparaging remarks about their competitor in their manual, I have never had problems with Curtis potboxes and I prefer the prepackaged assembly. The one shown in Figure 1.1 is an earlier model of the Curtis. The wiring shown in **Drawing 7** is a little unconventional in that Evnetics accelerators are intended to take a +5V and ground input across their 5K pots and produce a 0-5Vsignal to the throttle input on the controller. This requires three wires from the control enclosure all the way through to the cab. The wiring shown taking power from the 12V input through a 7K resistor requires only two wires and works fine.

The 12 V supply is taken from one of the 12V receptacles on the dashboard. If your truck has a switchable receptacle, it will save you from having to switch off to avoid running down the truck battery at night. Otherwise buy a switchable plug with an indicator light and at least three feet of wire.

The 12 V supply goes to the junction box. This can be a simple Radio Shack plastic project box with the aluminum cover sawn in half and turned outwards so you can drive sheet metal screws through it to mount it. Inside the box is a pair of eight-position terminal strips for control and instrument connections. The first two positions on the control strip are +12V and Ground from the dashboard receptacle which are jumpered to two positions on the instrument strip. The +12 V is connected to the red wire in the four-conductor cable to the motor box. A 7KΩ resistor goes from the +12V to a terminal then to the white wire from the accelerator and the white wire to the motor box. The black wire from the accelerator comes back and connects to ground and the black wire to the motor box. Check to be sure that between black and white there is a 0-5000 Ω resistance as the accelerator pedal is depressed. The microswitch wires from the accelerator, usually blue and brown, connect the +12 V to the green wire going to the control box. The red wire turns on the primary contactor and the controller when 12V is applied. The green wire turns on the secondary and motor contactors when the throttle is depressed. The white wire provides 0-5V to the controller to control speed. The black wire is control ground and signal ground.

The instrument wiring on RJ cables passes through the junction box and up onto the dashboard through black plastic ribbed split loom. If the anchoring plastic clips are painted black to match the loom and the dashboard, the whole connection is not too noticeable. The 12 V switched supply plug on the dash provides a midpoint anchor. The RJ-25 cable from the battery loop simply passes through the junction box and plugs directly into the Manzanita display, which is mounted on the dash with double-sided tape.

The instrument cable is interrupted in the junction box by connecting two conductors to +12 V and ground on the instrument strip. The other conductors are battery voltage (output 1 on the Evnetics controller), battery current (output 2), motor current (output 3) and overtemperature ground from the motors. They connect to meters on the dashboard shown in Figure 3.10. Two or three standard 2" 50 mV full scale meters can be used. The ground is common to all of them The volt meter can be set up with a suppressed ground at the controller via the web site interface so that it reads 150 to 250 V if desired. The current meters can be separate or a single meter can be switched from one to the other as shown. The +12V goes to the overtemperature lamp which is normally on for Normally Connected Warp motors and normally off for Normally Off ADC motors. The +12V and ground also power the instrument lamps and the display as shown in **Drawing 7**.

Figure 3.10 Instrument cluster showing the analog voltmeter and the ammeter which can be switched for reading battery or motor current. The Manzanita display shows all cells but one fully charged and the temperatures (yellow line) low.

Options

Single Exhaust

The minimum that is required to make space for the conversion is to redirect the exhaust pipe by 90 degrees and run it under the right frame rail as shown in **Drawing 2**. At that point it is turned 90 degrees to the rear and either fed to a muffler suspended under the door, exhausting just ahead of the rear wheels or turned again, brought back under the frame and reconnected to the existing muffler. The choice may depend on the age and condition of the existing muffler and tail pipe. If old, replace. A new heat shield for the crossover pipe is required per **Drawing 2** and has to be removed each time the drive is accessed for inspection or maintenance. Put two 2 ½ inch muffler clamps on the crossover pipe in the right place to hold the new heat shield and secure it with additional nuts on the clamps. Once you have them adjusted properly, the clamps need never come off when you remove the new heat shield.

Valence and Lead-Acid Batteries

Lithium-ion batteries from Valence Technology, Austin, TX, are designed to be drop-in replacements for Group 24 and Group 27 lead-acid batteries, and can fit in the same mounts. They are sealed, weather-tight units which can withstand the wet and dirty environment in the truck bed with the bulk of the battery management electronics built in and protected in each battery.

The weight of a 12.8 V group 27 Valence battery is 43 lb. Two pairs can be mounted together to provide a nominal 48 V in the same footprint as four group 27 lead-acid batteries with more capacity and half the weight. A complete module with batteries weighs more than can be conveniently be handled by one person, but the battery mounts can be installed first and the individual batteries installed later constituting a single rigid unit. A suitable battery mount design is shown in **Drawing 4,** along with the drilling template for the truck bed.

After the battery installation and wiring, the control enclosure is installed. For Valence and lead-acid batteries the enclosure is a modified cross bed tool box shown in Figure 3.11 and **Drawing 6.** The enclosure is preassembled with the controller, the charger and the contactors and connectors to complete the pack wiring. Four Anderson 350 Amp connectors are used to connect to the four 50 V battery modules at the front of the box allowing the electronics enclosure to be removed without disconnecting the batteries. These connections are covered by a 9" wide sheet metal spacer between the front of the truck bed and the front (hinged) side of the enclosure.

A standard cross bed tool box, ten inches deep by twenty inches long running the full width of the truck bed will be sufficient for the enclosure, and will cost around $250. Details of locating the power electronic components in the box are shown in **Drawing 6.** The box will need four oblong holes cut in the lower front (hinged) side for the Anderson connectors. This can be done with a 2" hole saw and a saber saw with a metal cutting blade for the straight cuts between holes. The openings should be flush with the bottom of the box to promote drainage of any water that gets in. You will also need two 2½" holes for cooling the charger, two more if you use the Evnetics controller and one or two for the 2" conduit connection(s) to the motor(s). The hole pattern and dimensions are shown in the drawing.

Figure 3.11 Cross bed toolbox installation in an earlier truck.

The Anderson connectors are screwed into the bottom of the box at each of the oblong holes as shown in the plan view in **Drawing 6**. They may be simply fastened flush with the bottom, making sure that they are the right way up so that the connector can be assembled, or more elegantly, they can be mounted on standoff brackets to center them in the holes. The power electronic components are located in the bottom of the tool box as shown, The Curtis 1231 controller and the Zivan NG-3 charger are shown with mounting dimensions and connections as well as the wiring details of the Valence BMS. Dimensions and hole patterns for the Manzanita and Evnetics units can be copied from the lithium-ion enclosure in **Drawing 5** if desired instead of the Curtis and Zivan.

Control Box Assembly

Once the major components are mounted in the box, the wiring can begin according to Figure 3.4 (**Drawing 7**) for lead acid batteries or Figure 3.12 (**Drawing 7A**) for Valence batteries. The major difference is the Valence BMS unit which resides in the tool box and controls the charger and motor contactors to cut off the power if the batteries are in distress from over charging or over discharge. Make all the high current connections with welding cable and crimp lugs as described above. Then install the 240V AC connection to the charger and the 205 V DC connection from the charger to the controller. Finally the control and instrument wiring is added. All the high power wiring should be routed as directly as possible to minimize losses and the length of expensive cable. Keep the two motor cables as close to one another as possible to minimize RF interference from the PWM controller, (you will hear it on your AM radio). The low power wiring should be run along the rear of the box and be bundled neatly with numbers on each end of each wire to permit tracing later.

The power cables to the right side motor as well as the 12V control circuitry coming into the box and the instrument lines leaving it are all routed through a 2" conduit located as shown in **Drawing 6**. If the box has a shelf support kink formed in the front wall it will be necessary to use a flat plate or bracket riveted to the outside and bridging the kink to provide firm support for the 2" 90 degree elbow fitting (McMaster Carr 7119K96). The conduit section leading from the box to the side of the truck bed is 37 inches long. Liquid-tight conduit (McMaster Carr #8071K47) is more rugged, Ultra-flex (McMaster 8069K17) is easier to install. This conduit is installed on the box with all the wire running through it prior to installation of the box in the truck so that all the wiring is preassembled.

For series-wound traction motors, the conduit can also serve to conduct clean cooling air to the junction box on the right hand motor from the control box. If there are two motors, an empty conduit can be installed on the left side to cool the left motor. The motors draw in air at the brush end, which is covered by the junction boxes, and expel it from the drive end. The junction boxes on the motors shield the brushes from the wet and dirty environment under the truck.

Control Box Installation

Once the box is wired, it is installed as a unit along with the cabling to the motor and the control and instrument cables. The 37" section of 2"conduit leading from the control box should be fed through a 2" straight fitting and a 2"pipe coupling which are left loose. The whole assembly is placed on the truck bed and slid forward to leave enough room for the cabling to be fed through the hole in the side of the truck bed.

The outside 90° 2" elbow should be threaded onto the cabling and slid up from under the truck to poke through the 2 3/8" hole and mate with the 2"coupling which can now be tightened into it. The conduit is completed by screwing the straight fitting into the coupling, and finally tightening the straight fitting onto the 37" of 2"conduit.

At this point the box can be secured to the bed by the J bolts supplied with it and the aluminum diamond plate spacer added to cover the space between the box and the front of the bed, as shown in **Drawing 6**. The 37" conduit should allow enough slack so that the box can be pulled to the rear to inspect and service the batteries without disconnecting the wiring.

Lead-Acid Battery Installation

Lead-acid batteries must be separated from the electronics because of corrosion. Only flooded (and vented) batteries are recommended for this service because sealed lead-acid batteries cannot stand the high, continuous current drain required. When charging, flooded batteries emit a fine fume of sulfuric acid droplets originating when hydrogen gas bubbles rise to the surface of the electrolyte and break. This fume will go everywhere and corrode whatever it touches. It is a must to cover the tops of lead acid batteries, particularly around the vents and the positive terminals with baking soda to neutralize as much as possible of the acid. This in turn means that the batteries must be accessible to renew the soda. Access is necessary on a monthly basis to add water to replace that electrolyzed away during charging, and to tighten the terminal connections. The soft lead will not maintain tight connections and tight connections are critical in these high current applications. The author has melted several battery terminals due to laziness in this regard.

The most convenient way to mount lead-acid batteries is in battery mounts screwed to the floor of the truck bed per **Drawing 4,** with the electronics in a cross bed tool box above them as described above for Valence lithium-ion batteries. This provides a secure but isolated mounting for the batteries and an inexpensive and compact way to provide a weather-resistant, damage-resistant environment for the electronic components. It places them immediately above the batteries and with a skirt below and a spacer ahead it shields the batteries from the weather as well. A battery mount design suitable for Trojan 1275 batteries is shown in **Drawing 4** along with a drilling template for drilling the mounting holes in the truck bed.

A remote watering system is a very useful accessory for a flooded battery system that needs water monthly. We have used a system that manifolded four batteries with 24 individual cells to a single watering point. The system had valves in each of the cell vents which let in just enough water to fill each one from a hand operated bulb pump. Three systems watered the twelve batteries of a 144V pack taking about a gallon per month of daily driving.

Zivan Charger

A Zivan NG-3 charger is shown located on the left side of the box in **Drawing 6**. It takes 230 or 240V single phase AC and delivers 3 kW of DC power at voltages up to 200 V and above. It is a switching power supply and can be programmed to deliver a wide variety of charging profiles. Programming requires that the charger be sent back to Electric Conversions (Elcon) in Sacramento, CA. The Zivan is air-cooled and a louvered and screened grille is needed as shown to provide positive exhaust of the cooling air outside the box Air inlet is provided for by the space around the Anderson connectors with a combined area of at least twelve square inches.

The charger is connected to the B+ and B- poles of the battery through a 35 Amp very fast acting ANN fuse. The current can be as high as 15 amps so #12 wire or heavier is indicated. (Note that fusing the output is indicated for a Zivan charger but <u>not</u> for a Manzanita.)

The main contactors on the battery bank must be closed to complete the series string for charging. This offers a route for the Valence battery management system to control the charging process by supplying 12 V to the Zivan charger contactor as shown in Figure 3.12. As before, the primary and secondary contactors can be powered by twelve volt power supplies powered from the 240 V input by taking 120 V from one line or the other and the neutral as shown in **Drawings 7 and 7A**. The neutral should be grounded to the box. The 12 V running power should be supplied through diodes on each contactor coil to isolate the charging control signal from the running control signal to avoid overloading the charging power supplies, and to allow charging to take place when the main 12V control power is off overnight.

Curtis 1231C Controller

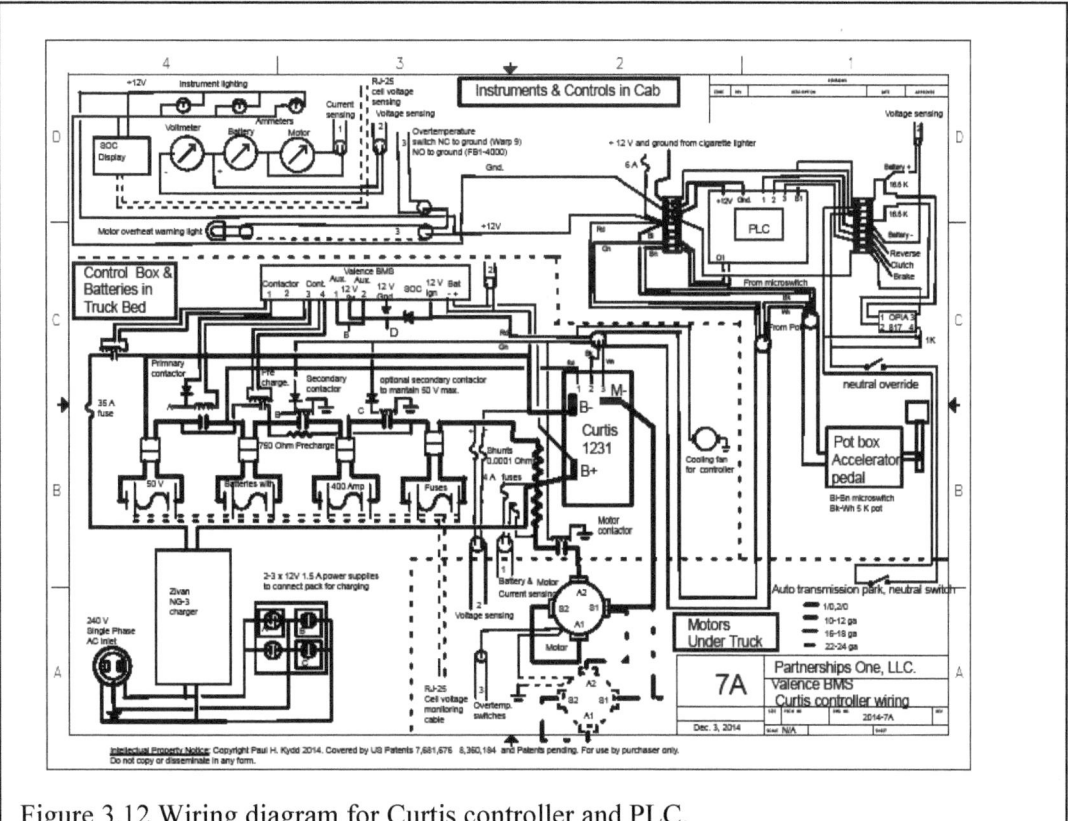

Figure 3.12 Wiring diagram for Curtis controller and PLC.

A Curtis 1231C- 86XX controller rated at 500 Amps peak and 144 V is shown mounted at the other end of the box in **Drawing 6** and Figure 3.12. It takes propulsion battery power directly from the end Anderson connectors on the main battery string to its B+ and B- terminals.. This controller can be uprated to take 200 V DC by David Mosher of Cedar Rapids, IA. Its nominal continuous current rating is 225 Amps. The current ratings can be increased to 1000 peak and 400 amps continuous.

The controller is air-cooled and requires a large, deeply-finned aluminum heat sink at least an inch and a quarter deep covering the entire flat bottom surface of the controller. The controller should be mounted to the heat sink with high thermal conductivity paste (McMaster Carr 3883K24) and screwed down to a rectangular hole in the bottom of the box through which the heat sink projects downwards. The heat sink should be provided with a 12 V DC fan operating from the 12 V control circuit so that it is on all the time the electric system is running. The fan should be mounted under the heat sink blowing air upwards through the fins.

If a Curtis controller is used with lithium-ion cells in the same enclosure, the heat sink should be mounted outside on the top of the lid directly over the controller and in good thermal contact with thermal paste. Because the heat sink is exposed to the slip stream when the truck is in motion, no cooling fan is required. The slip stream is actually across the truck bed behind the cab so the heat sink and the controller are mounted perpendicular to the direction of motion. When so mounted, the potentiometer adjustments for battery current limit, plug current limit and ramp rate are facing the edge of the lid for easy access.

The controller needs electronic power relative to battery ground which is taken from B+ through a relay operated by the 12V control system to pin 1 to turn it on. It also needs a 0-5000 ohm signal on two wires from the pot box to pins 2 and 3 to set the pulse width and thus the power taken from the batteries.

The controller should be set to limit the current to the motor to avoid excessive battery drain and excessive torque on the electric drive. For Curtis controllers this is done by removing the center socket head plug on the side of the controller and turning the potentiometer behind it counter clockwise with a small, insulated screwdriver. Another potentiometer allows for adjustment of the current ramp rate and a third limits plug braking current in which the motor is switched into reverse to slow fork lifts and other low speed vehicles, (not recommended for full size EVs).

The control wiring for Curtis controllers is different in that all the interlock functions performed by the Evnetics controller must now be performed by a separate Programmable Logic Controller (PLC). This requires a four-conductor 18 gage cable to carry the signals from the transmission and brake switches to the motor box as before, but an eight-conductor cable to carry the red and green wires from the control enclosure to the 12 V power connection in the truck cab and the black and white wires to the pot box, plus the three wires for brake, transmission and reverse from under the truck to the PLC. This cable should originate at an eight-pin female receptacle on the back wall of the motor box (McMaster Carr #3216K61) which is connected to the PLC by a corresponding cable (McMaster Carr #3216K31) running up through the hole behind the right seat as before.

The instrument wiring is also different than for the Evnetics controller because the Curtis does not provide voltage and current signals. Thus the instrument wiring must be fully analog and rated for 200 V DC. The instrument leads must be fused in the control box since they are connected directly to the battery. Four conductors are required to get the instrument signals to the motor box and on to the PLC. Another conductor is required for the overtemperature switch output from the motors to the motor box and on to the PLC. The cleanest way to do this is to use a five conductor cable from a receptacle on the motor box (McMaster Carr#3214K36) and corresponding cable (#3214K34). The instrument cable is passed up through the hole in the floor along with the control cable. This completes the work under the truck.

When connecting the high current cables to the Curtis controller, use two wrenches to avoid imposing torque on the buss bars in the controller which may damage the internal connections. Screws of 5/16" diameter (8 mm) are minimum and should be tightened to 15 ft lb. It is necessary to retighten them every once in a while to ensure that they have not loosened

Installing the PLC and Electric Accelerator Pedal

The main node of the Curtis control system is the Programmable Logic Controller (PLC) which should be mounted on the transmission hump in the cab. It takes 12V DC power from the cigarette lighter receptacle on the dash to a terminal strip on the left as shown in the wiring diagram. A 12 V plug with a built in switch and indicator light is useful so that the electric drive can be turned on and off at will. The plus and minus 12 V connections are jumpered to adjacent terminals to provide for the number of connections to be made.

The first of these is to the PLC itself. The Curtis accelerator pedal is mounted on the floor just behind the IC engine accelerator pedal as shown in Figure 1.1. The blue and brown wires from the microswitch in the Curtis connect the +12V supply to output relay O1 on the PLC. The red wire in the 4 wire cable to the control box via the motor box is connected to the +12 V and closes the primary contactor as soon as 12 V power is available. The green wire takes the output from O1 on the PLC and energizes the secondary contactors and the motor controller once the PLC is satisfied and the accelerator is pressed down.

The inputs to the PLC are on the right hand terminal strip. High voltage DC from instrument cable 2 is fed to a voltage divider generating an ungrounded six volt signal which is fed to opto-isolator OPIA 817. The output photodiode of the isolator provides a grounded signal of a magnitude that can be adjusted by the 1 KΩ potentiometer. This signal is fed to terminal B1 of the PLC and adjusted so that the PLC will open relay O1 if the battery voltage falls below the desired minimum.

The other inputs are digital on/off +12Vsignals derived from the vehicle. The first of these is the reverse signal taken from the backup light switch on the transmission. It prevents the electric drive from trying to go forward when the IC engine is in reverse and both throttles are pressed.

The next signal is from the Park or Neutral positions on the automatic trans-mission. The transmission provides a switch that enables the starter only in Park and Neutral. This switch is used to short the opto isolator and disable the electric drive. It prevents the electric system from operating when the vehicle is immobilized in park and allows the IC engine to be started and warmed up without moving. The same function can be provided by the clutch position switch on manual shift vehicles providing a +12V signal to the second digital input of the PLC. The park-neutral switch connection needs to be opened with a manual override switch to permit electric operation in neutral for automatic transmissions without free wheeling.

The final input is the brake signal taken from the brake light switch. It prevents the electric drive from trying to propel the vehicle when the brakes are on. The software supplied with the PLC requires all of the inputs to be in the affirmative before closing relay O1 and permitting the electric drive to operate. Other functions such as over temperature protection can be added to the PLC.

Programming the PLC

The only PLC used on this project has been the Crouzet Millenium 3. It can be programmed using ladder language or Functional Block Diagram language. The latter is more intuitive, as to what is being simulated and has been used. The program is shown in Figure 3.13.

The three 12 V signals for reverse engaged, park or neutral engaged (IC engine start) and

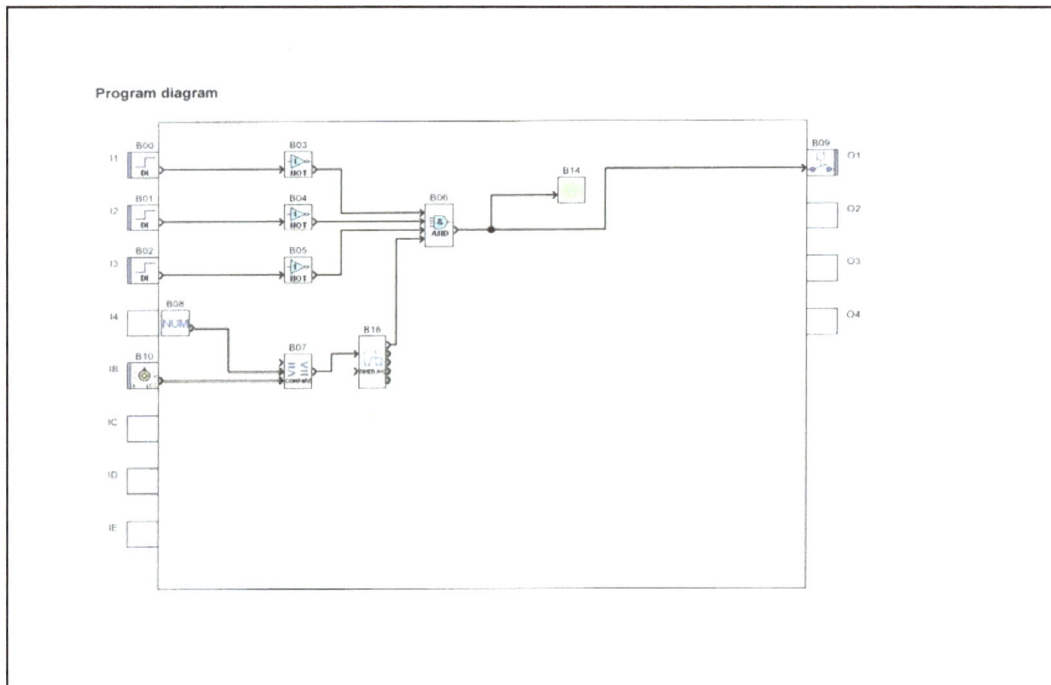

Figure 3.13 Program CAR to interlock the controller against operation in reverse, while braking, while starting the engine or with low battery voltage.

braking are input on I1-I3, Blocks 00-02, which are digital inputs with +12V signaling active. These inputs are inverted in Boolean NOT nodes, Blocks B03-05 so that zero volts is active and summed in the AND gate B06. The signal from the opto-isolator monitoring the battery pack voltage is input through a potentiometer at Block B10 which outputs a number between 0 and 1023 proportional to the input voltage between 0 and the 12 V supply to the PLC. This number is compared to a preset number set at 450 in the comparator. Nominally the output of the opto-isolator is set at 6V when the pack is at rated voltage of 3.2V per cell for Valence or CALB batteries resulting in an output from the potentiometer of 512. A value of 450 thus corresponds to 88% of nominal voltage or 2.8 V/cell which outputs an inactive state to cause shut down. After a delay of 20 seconds imposed by the timer to eliminate momentary voltage dips due to acceleration, this signal is transmitted to the and gate and on to the output relay on O1.

If the battery is not too low the system will restart if the voltage becomes high enough to turn the comparator on again. A latching function can be built in to the program to prevent this. Instructions for programming the PLC are provided with it along with extensive on-line help.

Dashboard Instruments

The dashboard instrument package is mounted on the dash board immediately in front of the driver with double sided adhesive tape. It comprises an ammeter to display the current drawn from the battery and thus the power being delivered to the drive unit. A second ammeter may be added to measure the motor current, (which can be higher than the battery current), if desired. The ammeters are connected directly to shunts in the control box via instrument cables 1 and 2. There is a volt meter or a state of charge meter to measure the capacity remaining it the battery. The volt meter is connected to a continuation of instrument cable 2 from the PLC. More elaborate displays such as that from Manzanita Micro will require a special cable from the BMS through a connector in the motor box to the PLC and on to the instrument cluster.

Instrument cable 3 operates the over temperature warning light. Advanced DC motors have a normally open over temperature switch. Vehicle ground is connected through instrument cable 3 to the motor box, the over temperature switches in parallel, and back through the PLC enclosure to the instrument cluster where it energizes the over temperature light. Netgain Warp motors have a normally closed switch. In this case +12 V is sent through instrument cable 3 to the motor box the OT switches in series and back through the PLC enclosure to the instrument cluster where it energizes the over temp-erature light. The light serves as a power on indicator as well as an over temperature signal. When the motor is cool and +12 V is supplied, the light comes on indicating that the system is ready to function. If the motor overheats, the propulsion system should be shut down with the motor still attached to the drive shaft and turning. This continues the flow of cooling air through the motor drawn in by its own fan which will cool it rapidly.

The Twin 9" Motor Drive

Drawings 13 through 17 show the drive required to use conventional radial-flux series-wound DC traction motor such as the Warp 9 and the ADC FB1-4001. While these motors have provided adequate performance and are slightly less expensive than the Agni 11" they suffer from much greater weight and length. The radial flux motors weigh three times the axial-flux motors for the same power output and the total drive assembly weighs four hundred pounds rather than two hundred, making for more difficult installation and poorer vehicle performance and payload. Because of their length the radial flux motors interfere with a cross member under the floor of the truck and have to be mounted two inches lower than the drive shaft centerline. To balance the forces on the shaft this requires smooth faced idlers bearing on the back side of the Poly Chains, which causes noticeable wear on the back side and probably shortens chain life. These motors also have undesirably short drive shafts which makes secure mounting of the drive sprockets difficult.

Drawing 13 shows the layout of the twin motor drive with the low centerline and the smooth faced idlers. These are made by turning off the teeth of regular idlers as shown in **Drawing 17** while retaining the mounting screws, modified as shown. The idlers should be mounted by screwing them into the ¾-16 nuts welded to the motor mount brackets shown in **Drawing 16** with plenty of removable (blue) Locktite. They should also be secured with a 3/32" steel safety rod between the holes in the heads of the two heavy screws holding the idlers. In operation the rotation is tending to unscrew these screws, and if one came loose it would be a disaster. The good feature of the smooth faced idlers is that they can be installed last which makes installation of the Poly Chains over the flanged idler and drive sprockets easier.

The base of the drive is fabricated as shown in **Drawing 16** very similarly to the twin Agni drive. The idlers and idler support post shown in **Drawing 17** are identical to those of the Agni. The difference is the 2" offset of the motors to the drive shaft and the different motor mounting holes in the motor mounting brackets. The driven sprocket is now 80 tooth instead of 67 tooth but otherwise the same. The electric connection boxes shown in **Drawing 17** are very different.

Before installation the motors should be adjusted for clockwise rotation as viewed from the front. The brush assembly can be rotated by removing the axial screws that hold the rear bell on the motor and rotating it clockwise as viewed from the rear to advance the brushes against the direction of rotation.

Figure 3.14 Twin Warp 9 motor drive with installation cradle.

The motors are installed as before with 3/8-16 grade 5 cap screws and centered by the 4" holes in the motor mount brackets. The drive should be pre-assembled because the motors are so heavy that it is very hard to install them and get them properly aligned once the drive is in the vehicle. Once the motors are installed, the idler assemblies are mounted on studs which are welded flush on the brackets to allow the motor shafts to extend through the brackets as far as possible. The studs are lubricated with lithium grease and the idler assemblies and balancing springs are slid over them. The springs on the lower idlers are wound up one turn and slipped through the holes in the mounting brackets as before. The idlers are supported by a triangular plate at the outer end and a post which should now be installed with 12 mm x 100 mm cap screws and self locking nuts. The ends of the idler studs are secured with thin ¾-16 nylon insert lock nuts making a rigid but light assembly.

Slip the taper lock bushings onto the motor shafts with the 40 tooth drive sprockets and tighten the bushings to line the sprockets up with the idlers. The bushings should be installed with ¼" square keys and provision to ensure that the key cannot fall out such as thread locker or staking them in place. The idler tensioning cable assemblies are made up as described above for the twin Agni drive.

The final assembly is to bolt the electrical boxes on the back of the motors because it is difficult to install them once the drive is in the truck. Leave the covers off for now to allow for wiring once the drive is installed. A wooden cradle under the motors and snug against the side of the formed crossbar as shown in Figure 3.13 will allow the assembled drive to sit level on a transmission jack, and is highly desirable to simplify installation and make it safer. The entire unit weighs close to 400 lb, and it should be secured to the jack with a chain or nylon strap. This finishes the assembly of the drive.

Installation of the Twin Drive

Start the installation as before by removing the drive shaft and installing the 80 tooth driven sprocket as described for the Agni drive. Loop the 1200 mm Poly Chains over the transmission and reinstall the drive shaft.

Because the twin drive is so long and heavy it needs support at the rear end as well as under the sprockets. The 2" channel shown in **Drawing 26** and two pairs of 1" x 16 ga. galvanized steel straps are needed for this. It is important to have all the parts ready to go before installation because all are needed before you can take away the jack.

Jack the drive unit up under the truck and carefully align the idler sprockets with the 80 tooth driven sprocket on the drive shaft. The right side motor support bracket and the right hand electrical box should be snug against the right hand frame rail. Once aligned, clamp the drive to the frame with the clips shown in **Drawing 16**. Support the rear ends of the motors with the channel iron cross member and straps. The motors should parallel the engine and transmission by tilting down toward the rear by 4.5 degrees. It will be necessary to drill only the tops of the straps before assembly and mark and drill the lower holes once the motors are in place.

Once the mechanical mounting is complete the jack can be removed and the 1200 mm Poly Chains can be brought back over the drive shaft sprocket and the flanged idlers. Lastly install the smooth rear-side idlers with Locktite and the 3/32" tie wire and tighten the turnbuckle to tension the chains as before.

Bring the series power cabling to the motors as before and add the control and instrument connections through the right hand electrical box. The left hand box is connected to the left hand 2"conduit coming down from the truck bed to supply clean cooling air. The positive cable from the controller should be red and will need about nine inches of extension beyond the end of the conduit to connect to terminal A2 on the motor, which is inside the box on the top right hand side of the right side motor, where the conduit comes in. The negative cable from B- should be black and extend approximately 40 inches to come out the front of the box and connect to terminal S1 on the top of the left hand motor. You will need a separate short jumper cable of either color about sixteen inches long to connect A1 to S2 on both left and right motors. A 40 inch red cable connects S1 on the right motor to A2 on the left motor to complete the series connection to both motors. To reduce RF interference from the PWM controller try to run the motor cables as close together as possible. The wiring is easily disconnected for removal of the motors and the drive by undoing the welding cable connections to A2 on the right motor and S1 on the left motor and withdrawing them through the hole of the 2"conduit.

Figure 3.15 Single motor drive, rear side, with electric clutch plate and driving plate.

The Single Motor Drive

The assembly for the single motor drive unit is shown in **Drawings 18-26** and from the rear in Figure 3.15.

The motor drives the shaft extension visible on the lower right through an electric clutch shown disassembled with the driven plate on the drive and the magnet and driving plate on the bench. In later modifications the clutch has been replaced by a rigid coupling as mentioned in Chapter 1, and the rear ball bearing shown has been eliminated in favor of a single bearing on the front face. The motor is spaced from the face of the Poly Chain unit by a two sided spacer box shown in **Drawing 25** that provides a rigid connection and positive alignment but allows access to the Poly Chain from below and the left (drive shaft) side. The Poly Chain drives a companion sprocket on the upper left to provide torque to the drive shaft from both sides. Two idlers tension the chain and allow for differential rotation of the driven sprockets from the driver. The idlers are tensioned by a turnbuckle, a spring and cables partially obscured by the left idler sprocket

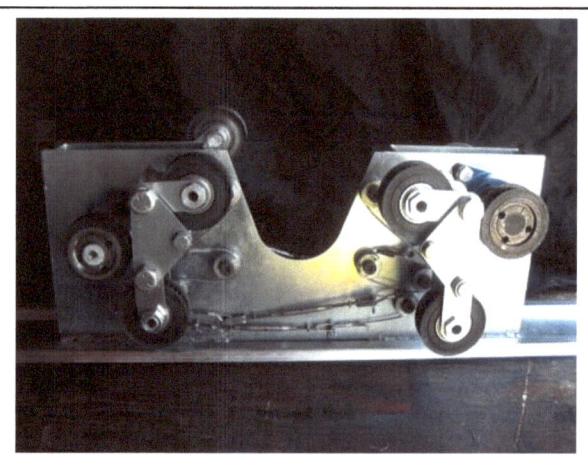

Figure 3.16 Single motor drive unit from the front showing idlers and drive sprockets.

The same unit is shown from the front in Figure 3.16. The two drive shafts drive sprockets with the motor driven sprocket now on the lower left. There are two sets of idlers tensioned by turnbuckles and cables as before such that one turnbuckle tensions the lower left and upper right idler and the other tensions the upper left and lower right. This arrangement has now been replaced by loops of cable on each side tensioned by a single turnbuckle as mentioned earlier relative to the twin Agni drive. The vehicle drive shaft with the large driven sprockets mounted on it fits into the notch at the center of the drive as shown in Figure 3.17.

The drive is mounted on a formed steel cross member with bolt holes mating with clips that secure it to the frame rails on both sides as shown in **Drawing 21**. As with the twin installation, it is important to avoid drilling into the frame, which is a carefully designed high-strength assembly with limited tolerance for damage, particularly to the flanges top and bottom.

Assembly of Drive

Begin by fastening the spacer box of **Drawing 25** to the Warp 11 motor with four 3/8 -16 grade 5 screws and lock washers (or the more complex fastening shown in **Drawing 25** for the AC Propulsion motor). The motor is automatically aligned with the box by the close-fitting rabbeted 4" hole. Note that the length of the spacer box is different for the two motors.

Next assemble the motor shaft to the motor with the differential tensioner shown in **Drawing 23** leaving a 3/8" space between the driving shaft and the shaft on the motor to allow for future cross chain installation and removal. Loosely assemble the 40 tooth Poly Chain sprocket for the cross chain with its taper-lock bushing on the tensioner. The socket head screws in the bushing must be facing out away from the motor. The motor and spacer box are now assembled to the drive unit with the shaft extending through the bearing holes on the front and rear sides of the assembly. Use 3/8-16 grade 5 cap screws with plenty of Locktite thread locker and lock washers. The dowel pins will ensure accurate alignment. Support the motor on wooden blocks or a cradle during assembly to ensure that its 225 lb weight doesn't distort the drive. Install the mounted 1" bore ball bearing on the free end of the motor shaft and screw it to the drive unit with ½- 20 cap screws and lock washers. Loosely install the 34 tooth Poly Chain sprocket on the free end of the shaft with the socket head screws pointed forward for access. Loosely install the counter shaft on the other side of the drive with two mounted 1" ball bearings and a 40 tooth sprocket aft and a 34 tooth sprocket forward to match the motor installation.

Next the idler assemblies are mounted on the drive assembly with long ¾-16 studs. Each stud has a ¾-16 jam nut tightened down onto the welded-in nuts on the drive with plenty of Locktite on both the lock nuts and the weld nuts leaving the rest of the stud free to allow the idler assembly to rotate freely on it. The rear side idlers have long levers and are assembled so that when the levers are both pulled to the left by a cable running over a pulley mounted to the left, the idlers move outward to tension the cross chain. Run the low height nylon insert locknuts onto the ¾" studs just enough to minimize clearance but allow the idler assembly to swing freely. If overrunning clutches are installed in the idler assemblies to provide automatic adjustment the assembly procedure is the same but the clutch arms are not fastened to the drive until the last step of the adjustment process after installation in the truck. Alternatively the idlers can be supported under load by self-adjusting lever and rack assemblies shown in **Drawing 22** which are engaged at the very end when the drive is in the truck and all the other adjustments have been made.

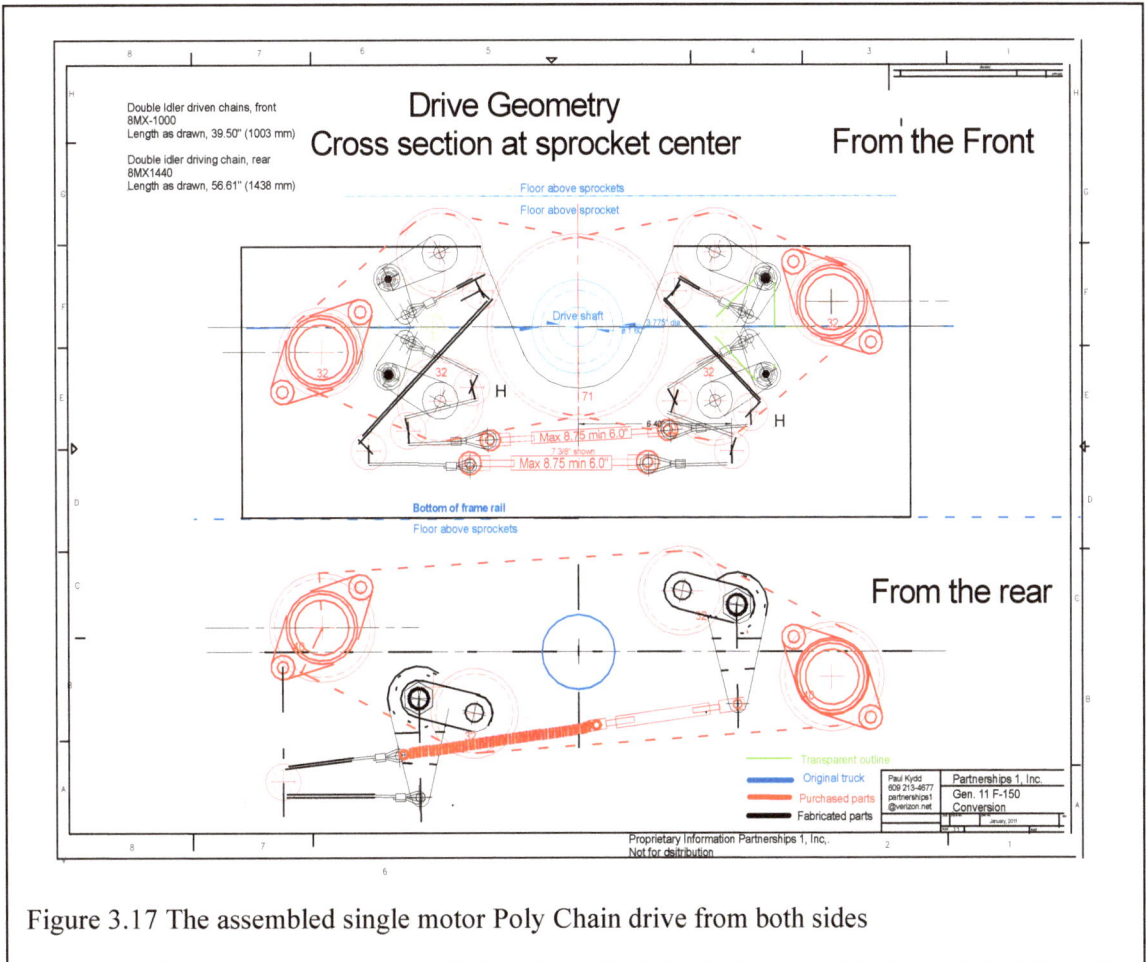

Figure 3.17 The assembled single motor Poly Chain drive from both sides

It is now possible to locate the 40 tooth sprocket on the counter shaft to line up with the idlers and the 40 tooth sprocket on the differential tensioner. Align the differential tensioner sprocket with the idlers and lock it to the motor shaft with set screws. Align the counter shaft sprocket with the idlers with a straight edge and lock it in place with the socket head screws in the Taper-Lock bushings. The bushings should be installed with ¼" square keys and provision to ensure that the key cannot fall out such as thread locker or staking them in place.

The idlers are tensioned by a cable running over a single pulley as shown in Figure 3.15. The pulley should be installed with a 3/8" diameter x 5/8" long shoulder screw. The 3/32"cable should be preassembled to the idler with a thimble and a Nicopress sleeve. It runs under the pulley and up to a spring which attaches it to the other idler lever and tensions both idlers.

The two sets of idlers on the front of the drive assembly can now be installed. Note that the left set for the counter shaft are long and the right set for the motor shaft is short because there are two sprockets on the drive shaft, one behind the other. Install these idlers the same way as the rear ones except that they are tied together by triangular plates and spacer tubes to make a stiffer installation. Insert the ¾-16 studs with their lock nuts noting that the lower lock nuts are round and tightened by a pipe wrench.

The two lower idlers are held up by torsion springs, one right handed and one left handed which are installed loose over the locknuts now. The purpose is to counterbalance the weight of the idler assemblies so make sure that they are positioned to do that. The idler brackets can now be installed with the idler support posts and their 12 mm screws so that the whole triangular idler assembly comes down flush and tight to the drive. Once the idlers are installed, the 34 tooth sprockets on the driving shaft and the counter shaft may be installed in line with the idlers with keys and Locktite as on the rear. The last step is to hook the torsion springs onto the levers on the lower idlers and wind them up enough to pop the other end into the holes on the drive assembly. The springs will be snug around the lock nuts when properly installed.

Drawing18 shows a more complex arrangement of cable pulleys to tension the idlers than was used in later versions of the drive. However, the method of making up the cables and adjusting them is the same as described above for the Agni drive. The single turnbuckle approach in the drawing is better than the two turnbuckles shown in Figures 3.16 and 3.17.

The final assembly is to bolt the electrical box on the back of the motor. Leave its covers off for now to allow for wiring once the drive is installed. A wooden cradle under the motor and drive which allows it to sit level on a transmission jack is highly desirable to simplify installation and make it safer. The entire unit weighs close to 300 lb, and it should be secured to the jack with a chain or nylon strap.

Installation of the Single Motor Drive

Installation is essentially the same as for the twin motor drive described above. The main difference is the method of tensioning the Poly Chains. Install the three Poly Chains by pulling the front one meter chains through the gap between the drive assembly and the drive shaft. Loop them over the main sprockets and work them over the idlers and the drive sprockets. Tighten the turnbuckle making sure that the chains are engaging the sprockets fully and evenly and that the idlers clear the idler support posts for the triangular stiffener plate uniformly top and bottom. The springs on the lower idlers should support the weight of both top and bottom idlers to make this possible.

Once the front Poly Chains are properly adjusted and tight enough that they deflect only one chain thickness under a ten pound downward or upward load and everything is uniformly spaced to provide running clearance and balanced forces on the driven sprockets the rear Poly Chain can be adjusted so that it also is engaging its 40 tooth sprockets. The spring loading of the rear chain tensions it but allows it to conform to the position of the front chains which are locked in rotation by the drive shaft sprockets and the need to locate the front idlers evenly relative to the idler support posts.

Finally, the self-adjusting levers can be engaged with the racks to lock the position of the rear idlers under load but allow the locking position to creep forward as the chains stretch and the sprockets wear. They perform similarly to self adjusting brakes. If the rear idlers are equipped with overrunning clutches to perform the self-adjusting function, run the locking levers around counterclockwise and secure with ¼-28 screws to position them initially. The mechanical installation is now complete. The electrical wiring is essentially the same as for the twin drives except for the AC Propulsion drive which has its own wiring harness to connect the motor to the Power Electronics Unit.

The Light Twin Drive

Drawings 27-29 show a twin motor drive with 7" diameter series-0wound motors. This drive is not recommended for full size pickups and vans because of inadequate power. It could be used on older passenger cars with ladder frames and on light pickups with adequate space underneath. Installation is substantially the same as for the big twin drive.

Chapter 4. Start up and Troubleshooting

Start Up

Electrical Inspection

While the truck is still on the lift, inspect the electrical installation. Make sure that the power wiring is going to the correct motor terminals, and that all the terminals are tight. Make sure that all the instrument and control wiring is connected correctly, and that all the connections are tight. It is a nuisance to put the truck back on the lift to fix errors later. When you are sure that everything is correct, close up the covers on the motor box(s) and ensure that all the conductors are safely inside and not pinched or chafing.

When the entire electrical system is complete, inspect all the wiring in the power electronics enclosure and the truck cab to ensure that all the wiring is done correctly and all the connections are tight.

Mechanical Inspection

After finishing the work installing the drive under the truck, take time to ensure that it is installed correctly. Inspect the Poly Chains for alignment and tension. Ensure that all fastenings have been tightened to the proper torque including especially those on the drive shaft. Ensure that nothing is loose and that wiring will not chafe, both the newly installed wiring for the conversion and the original wiring on the truck that may be close to the newly installed parts. It is wise to rotate the drive shaft with the truck still on the lift and even run it on gasoline to make sure that the new parts are functioning properly and that the Poly Chains track near the center of the sprockets.

Then take the truck down from the lift and road test it immediately on gasoline, while you still have access to the lift. Listen for excessive noise from the chains and any other unexpected noise or vibration. An audible whine from the chains is normal but louder noise as from interference between the crossover chain and the driving chains of a single motor drive needs attention. Put the truck on the lift again and inspect the chains for tracking and wear. The chains should be running centered on the sprockets with clearance on both sides of all sprockets if possible. If not, at least one sprocket should be centered. If not, adjust the alignment of the entire drive by loosening the mounting clips and moving the entire drive in the indicated direction.

Redo the tightening of the driving chains and the adjustment of the cams on the crossover chain to ensure that they are not fighting each other. At the end of the day you should have a smooth running connection between the drive shaft and the electric motor(s) with minimal noise from zero to 60 mph.

Battery Charging

Before running the conversion it is necessary to recharge the batteries. Lead-acid batteries loose their charge on standing and lithium-ion batteries are shipped only partly charged. Your charger should be set for the type of pack you have. For Zivan and ElCon chargers this is done at the distributor, Electric Conversions of Sacramento, CA. Tell them what batteries you are charging when you buy the charger, and they will load it with the proper profile.

For Manzanita chargers you can do it yourself. Follow the instructions that come with the charger. Be sure to connect the charger to the battery before turning on the AC power. The charger must have a DC load or it will burn out. Do not use an external fuse with a Manzanita charger. If the fuse blows, the charger will destroy itself. To adjust the voltage downwards turn the voltage trim control pot counter clockwise with the insulated screwdriver provided. If your charger has the very handy digital voltage control, set it to the desired constant voltage charging level. Turn the current control pot counter clockwise to zero. Turn the power on and turn up the current control. The ammeter should show current going into the battery. If not, and the yellow light is on indicating battery voltage at or above the set point, turn the voltage control clockwise to raise the voltage set point until the yellow light goes off. The batteries should now be charging. Monitor the voltage. As the battery voltage comes up, you will need to periodically increase the voltage set point by turning the trim pot clockwise to keep the yellow light off. When the batteries reach the desired constant voltage charge level at full current, turn the voltage trim pot counter clockwise until the yellow light goes on. The charger will now supply constant voltage at diminishing current. The blue light will begin to blink showing that the charger is in constant voltage mode. The length of time the charger will spend in this mode can be set by the timer switch. It should be long enough for the batteries to reach 2.35 V per cell for lead-acid and 3.6-3.8 V per cell for lithium-ion. DO NOT EXCEED 3.8 V PER CELL FOR LITHIUM-ION CELLS, AS IT WILL CAUSE PERMANENT DAMAGE. In fact charging to only 3.5 or 3.6 V will extend the life of your cells without materially reducing their capacity. The first few battery charge cycles will require close monitoring to ensure that everything is proceeding properly and that the equipment is functioning properly.

Programming the Controller

Set the controller to limit the motor current to 300 A. For the Curtis controller this is done by removing the center plug with a 1/8" Allen wrench on the left side of the controller looking at the connection end. The pot can be adjusted with an insulated screwdriver. 300 A will be near the center. Turn the pot counter clockwise and then increase the limit with a test run at each step until you have what you want.

For the Evnetics controller the limit is set by connecting a laptop to the controller with an Ethernet cable and contacting the controller as a website. Enter HTTP://169.254.0.1/ into the address bar of the laptop browser. You will see a web page that allows you to configure the controller.

At the top of the page you will see inputs for power mode, brake and tachometer. "Power" mode allows you to select between high (quiet) and low (performance) PWM frequency. "Brake" reduces power to zero when a brake light signal is applied to the "Brake" input on the terminal strip. "Tachometer" allows you to set the number of pulses out for a single pulse in to the "Tach" input to drive automotive type tachometers from a once per rev pickup on the electric drive.

You will want to set the inputs I1 and I2 for reverse, and clutch or automatic transmission to turn off the controller on positive 12 V signals from either input. Configure the two inputs as "throttle limit" and set it to 0%. You can set the throttle dead band and half throttle current to shape the throttle response if desired.

You will want to configure the outputs to give meter readings for battery voltage, battery current and motor current. These are PWM gauge drivers that give a variable current signal at 12 V. A 220 ohm load resistor and a 1N5231B Zehner diode in series between the output and ground rescales it to 0 to 5.1V and a 100/1 voltage divider, say 2200 over 22 ohms, scales it to a 50 mV full scale meter. The battery voltage meter can be set with a suppressed zero if desired.

Finally, you can set:

    Minimum battery voltage at no current

    Minimum battery voltage at maximum current

    Maximum battery current

    Maximum motor voltage

    Maximum motor current

    Maximum motor power

    Slew (Current ramp) rate

The critical settings are minimum battery voltage to prevent over discharge, (2.5 V per cell at max current and 3.0 V per cell at zero current for lithium-ion batteries), and Maximum motor current (300 A to limit motor torque and stress on the drive).

Electric Start Up

Start the IC engine as usual and turn on the 12V to the electric drive. The volt meter should register that you have DC power. With the Curtis controller the voltage will rise slowly as the precharge resistors allow the controller capacitors to charge. When the voltage gets high enough that the PLC is satisfied, the voltage should come up to a full 200 when you blip the electric throttle. When the PLC is satisfied, you will see highlighting around digital inputs 1, 2 and 3 for transmission, brake and reverse and analog input 1 for battery voltage. Output relay O1 will be highlighted showing that the output relay is closed. With the Evnetics controller all this is automatic.

With full DC voltage you can advance the electric accelerator pedal and you should see current flowing in the ammeters and feel the vehicle begin to move. Acceleration from a standing start under electric power alone is very weak, so you will want to add some IC engine power by pushing harder to engage the IC accelerator. With the electric accelerator fully depressed you should see the preset limit of 300 amps on the motor current, but only about 100 A or less of battery current. As the speed builds up, the battery current builds with it to the limit imposed by the controller. This limit could be set at around 150 A to keep the drain on Valence batteries within their maximum continuous duty limit and to avoid excessive torque on the electric drive. At 205 V and 150-160 Amps you should be able to maintain 60+ mph on the level. For better acceleration Valence batteries can deliver 300 A for 30 seconds.

Driving the conversion should be very similar to driving the unconverted truck. Once the truck is accelerated to the desired speed, the electric drive will maintain that speed on the level and up minor grades. Most of the mileage will be covered on electric power for as long as the battery power lasts with the IC engine called upon to accelerate away from a stop, for passing, and to climb steeper and longer hills.

When the batteries are nearing exhaustion you will notice a voltage sag when you call for more current. This sag is much more pronounced with lead-acid batteries than with lithium batteries. With lead-acid at the end you are driving gingerly to keep the voltage from plunging out of sight. With lithium you will be driving pretty much full bore to the end, but you will notice that you can no longer get full current. Eventually the sag will become great enough that the PLC or the Evnetics controller will shut down the electric drive. At this point you will be driving on gasoline as before the conversion for as long as it takes to complete the trip and start to recharge the batteries.

Recharging

The same comments as above on first charging will apply to recharging. You will want to monitor the first few charge cycles closely to ensure that you are getting a full charge but not overcharging. If all is well, recharging will be simply a matter of plugging in the vehicle after you are through with it for the day. The batteries should be kept charged as much as possible. Lead-acid batteries deteriorate by sulfation if left at less that full charge. Lithium-ion batteries do not suffer from capacity loss, but they only equalize when charging, so maintaining peak performance requires that they be kept charged and the charger plugged in as much as possible. In the winter time the equalization can warm the batteries and maintain their maximum capacity, offsetting the effect of cold weather.

Maintenance

Maintenance requirements for the conversion are minimal. If you have lead-acid batteries, you will need to replenish them with distilled water once a month or even oftener if you use the truck every day. With a one-point watering system on each group of four batteries this is not too onerous, and such a system will more than pay for itself in the time and aggravation of moving the electronics enclosure and dealing with 72 recalcitrant battery caps monthly. You will also need to replenish the sodium bicarbonate on the batteries and tighten the terminals at least every three months.

With lithium-ion batteries none of this is necessary. If you have a BMS which will keep them equalized, as you should, they will function unattended for years. We have had trouble with Valence batteries in a boat due to the CAN bus connectors not functioning. Cleaning them with contact cleaner solved the problem. The electronics for the individual cell BMS tends to be vulnerable and needs to be protected from the weather. Those connections also may need periodic cleaning.

The remainder of the electronics should be maintenance free except possibly cleaning off the heat sink on the controller and vacuuming out any debris that gets into the charger.

The Poly Chains feature no maintenance requirement, unlike conventional roller chain. When well adjusted they will go for a year or more without attention. It is advisable to inspect the chains every 5000 miles or six months and tighten as necessary. If they loosen, you will hear the drive getting noisy and perhaps vibrating. In extreme cases the chains will slip with a loud rapping noise under high current and high torque. Back off on the current immediately and readjust the chains as soon as possible. Slipping is not a good condition.

Inspect the chains for wear. Minor wear at the edges where the chains are in contact with the flanges of the sprockets is normal as is wearing off of the blue color on the contact surface of the chain. Gross wear, particularly asymmetric wear is a problem that should be addressed. Check that the sprockets are properly aligned and haven't shifted. Check that he idler assemblies are firmly supported and running true. Fix anything that seems wrong and replace the affected chain(s). A set of on line spare chains wrapped around the rear end of the transmission and tied up out of the way will save you from having to remove the drive shaft.

To readjust the chains proceed as when installing them initially. Tighten the turnbuckle maintaining even tension on the two driving chains until they are tight and the deflection is one chain thickness under a ten pound load. Then tighten the cross chain and reset the spring loaded cams to keep it tight on a single motor drive.

The bearings on the single motor drive should be lubricated with grease when you change the oil in the truck at 5000 mile intervals. The idler ball bearings are lubricated for life and need no attention. Replace the whole idler sprocket if they fail.

Nothing in the cab needs maintenance except for cleaning periodically to maintain its appearance.

Troubleshooting

Electrical

If you don't get a voltage indication when you turn on the 12 V follow this sequence:

1. Make sure you really have 12V. If your cigarette lighter plug has an indicator light this is very handy. Otherwise measure it with a meter. A good digital volt meter is a big help in troubleshooting.
2. Check in the electronics enclosure that you have 12 V on the primary contactor and that it is closing.
3. If you have no voltage the problem is either a blown fuse at the PLC end or in the red wire from the PLC to the primary contactor. If the fuse is blown, trace out why it blew before proceeding.
4. If the contactor is not closing you have a problem with the coil (which we have seen fail) or the contacts (which can also fail, particularly if they are not enclosed).
5. Replace the coil or clean the contacts as appropriate. You may have to replace the contactor, and spare components are handy to have.

If you have 12 V and primary contact but the secondary contactors won't close to give you full voltage proceed as follows:

1. Check that the microswitch on the electric accelerator is closing and providing 12 V to the PLC or 0-5 V to the Evnetics. If not fix it or replace it. Check that the controller itself is getting the recommended input power.
2. With an Evnetics controller monitor the web page for fault messages. Check all the inputs to see if spurious signals on reverse for example are preventing operation. Check settings to be sure that the limits you set are fulfilled, e.g. the battery voltage is high enough.
3. With a Curtis controller check the PLC. See that all the three digital and one analog inputs are highlighted. If one or more of the digital inputs are not highlighted, you will need to find out why. Measure the inputs to see it they indeed are at +12V. If so, find out why and fix it. Possibilities are shorted switches for reverse, brakes, or clutch.
4. If the battery voltage analog input is not highlighted check to see if the battery is indeed charged and delivering its stated voltage. If it is, the problem is in the optoisolation circuit that samples the battery voltage or an open switch on Park-Neutral. There is a gain adjustment which may be used to offset component drift. By checking voltages on the various components you will be able to identify the problem and fix it, perhaps most easily with a replacement for the whole circuit.
5. If the PLC inputs are OK check the output relay to see if it is functioning. If not it may be a PLC problem or a software problem. Try reloading the PLC software. If that doesn't work try reprogramming the PLC to use another of its relays. If that doesn't work, you need a new PLC.
6. If the PLC is providing 12 V from the relay, check to be sure that it gets through to the green wire in the electronics enclosure. If not fix it. If it does, follow the same procedure

as with the primary to verify that the secondary contactors are working or to fix them if not.

If you have the full 200 V but the motors won't run:

1. Check the output from the controller. The M-terminal should be pulled down toward B- when the accelerator is depressed. If it is not there is a problem in the controller
2. If there is voltage to the motors check the connections to the armatures and fields to ensure that all are tight. Rotate the motors to see if the brushes are skipping. Open the electric box on the motor and remove the brush cover to see if there is major damage inside. We have seen this too! Repair if possible or replace.

Mechanical

The most likely mechanical problems are loose chains producing noise and vibration as mentioned above under maintenance. In general noise comes from the driving chains and the cross chain being out of synch and fighting each other in single motor installations. If you notice increasing noise or vibration at certain speeds:

1. Check the tension on the driving chains. If loose, tighten as described above. If unduly worn, replace. You should get at least 12,000 miles on a set of chains. If you don't something is amiss in the drive and should be fixed.
2. One possibility is a frozen idler assembly. The idlers have to move to accommodate engine vibration. Be sure that all idler assemblies are free to rotate.
3. Another possibility is a missing key or a loose bushing on the driving sprockets. Make sure that they are tight and right.
4. Yet another possibility is a worn or damaged ball bearing. Shake the shafts to detect play and rotate to feel and listen for damage.
5. Even if the driving chains are tight they may be out of synch with the cross chain on a single motor drive. Readjust it.
6. If you hear a crackling sound like a high current arc, It may be a sign that one of the chains is breaking up. We have had this happen, and it seems to be the individual fibers in the chain breaking. If you hear this, check to see whether any of the chains show signs of damage. Replace them serially to locate the offender. (Of course it may **be** a high current arc, so inspect the entire electrical system. You may be able to see indications of a problem on the meters if it is electrical. If you see no such signs it is likely to be mechanical.)

Chapter 5. Making Major Components Yourself

Battery Mounts for Individual Lithium-Ion Cells

Battery mounts for prismatic lithium-ion cells to be located in the electronic enclosure of **Drawing** 5 are shown in **Drawing 3**. The dimensions of the individual cells are shown in the lower center. They are for cells from China Aeronautic Lithium Battery (CALB) with capacities of 100, 130, and 180 Amp hours, all at 3.2 V per cell. The Thunder Sky 90 AH battery, another popular brand, is almost identical in size to the CALB 100 Ah but 61 mm thick instead of 67 mm making the stack 9% thinner.

Because the batteries are relatively light, 1" (25 mm) x 1/8" angle iron is specified except for the front cross pieces where 1 1/4" x 1/8" angle is used to accommodate the step down from the ribs in the truck bed. The rectangular mounts are welded up with weld nuts (McMaster Carr 93975 A300) to secure 9" 3/8-16 cap screws (McMaster Carr 91236A654) to hold down the 100 Ah batteries and 11" screws (McMaster Carr 91236A657) for the 130 and 180 Ah cells. The weld nuts can be welded direct to the angle cross pieces and aligned vertically by running one of the screws down through the nut to a locating hole drilled in the cross piece.

Begin construction by cutting the lengthwise angle iron parts, noting that the dimensions inside the mounting are <u>inside</u> dimensions and the dimensions outside are <u>outside</u> dimensions. Cut and drill the crosswise angles including the holes which will allow for aligning the hold down weld nuts.

Assemble the mounts on a flat surface. Ensure that everything is square and of the proper size. It is highly desirable to have the actual cells in hand to ensure that they will fit before welding. Tack-weld the assemblies together at the tops of the angles, being careful not to build too big a bead that will prevent the cells from fitting in. Turn the assemblies over and weld the angles together at all the seams. TIG welding or gas welding is much preferred for this. Stick welding and MIG welding leave too big a bead, which then has to be ground off to allow the mounts to lie flat. Use rod very sparingly and try for a flat fusion weld with good penetration for strength. Turn the mounts right side up and weld along the sides up to the tack to add strength. Again aim for a fusion weld on the outside angles. Build up as little fillet as possible to avoid installation problems.

The hold downs are fabricated out of 1" x 1/8" fiberglass angle (McMaster Carr 8542K51). You will need four 10' lengths of angle for 100 and 180 Ah cells and three 10' and one 5' length of angle for 130 Ah cells. There are 1" angle iron end pieces crosswise on top mirroring the end pieces of the base and securing the fiberglass rails with 6 mm screws as shown. Note that the hold down is 20 mm wider overall than the base. This is to clear the large hex terminals on the top of the cells. When finished, the hold downs are secured to the bases with long cap screws and washers, as mentioned above. Because of the non corrosive environment with lithium ion batteries, the mounts and hold down ends can be simply painted or galvanized for corrosion protection and appearance.

Install the mounts in the truck bed as shown on the left of the drawing. You can use one of the mounts as a drilling template to locate the pilot holes for mounting screws based on the dimensions given. The strongest mounting will be achieved with 8 mm (5/16") screws with nuts and washers. Unfortunately this is very difficult to do in a number of places, since the gas tank is immediately below the bed. Furthermore, there is a cross member just beneath the rearmost screw locations that makes it virtually impossible to access from below. The best way here is to fasten the mounts with self tapping 5/16" (8mm) screws and washers, either 1" (25mm) 5/16" sheet metal screws (McMaster Carr 90054A415) or thread cutting 5/16-18 screws (McMaster Carr 90096A583). Position the mounts properly and drill ¼" (6mm) pilot holes <u>being sure not to drill into the gas tank.</u> Use a drill stop to make sure you do not commit this disaster. Then drive the screws from above securing the mounts firmly to the truck bed.

Battery Mounts For Lead-Acid and Valence Batteries

Details of battery mounts for batteries which do not need to be enclosed are shown in **Drawing 4 Lead-Acid—Valence battery mounts**. Plan view details are shown on the right for two types of mounts, the upper one suitable for Trojan 1275 flooded lead-acid batteries and the lower one suitable for Group 24 and Group 27 lithium-ion batteries from Valence Technology and for lead-acid batteries of these standard sizes. The truck bed elevation view shows single examples of the two types of mounts fitted to the truck.

The plan view on the left shows how the full sets of battery mounts fit into the truck with phantom details of the batteries. The 1275 batteries are shown at the top and the group 27 batteries below.

The two mounts are structurally similar but differ in details, due to the fact that the 1275 battery hold downs run across the tops of the batteries and are held down with J bolts while the Valence hold downs run lengthwise and are held down with cap screws. In both cases, the batteries are located by a welded frame of 3mm (1/8") angle iron bolted to the truck bed and resting on the ridges in the bottom of the bed. The Trojan mount uses 1 ¼" (32 mm) angle with a 1 ½" (38mm) cross piece at the front end. The Valence mount is similar with a 2" (51mm) cross channel in the center. At the front of the mount the crosswise angle iron is stepped down to lie flat on the bed and provide a solid connection, as shown in the detail.

For the Valence batteries you will have to cut notches in the lengthwise angles to locate the crosswise 2" (51mm) channel section which provides a center anchor for the hold downs. Drill the 1.75"x2" channel sections and the ends 3 5/64" (14) mm and weld in 3/8-16 weld nuts, (McMaster Carr 93975A300) which will anchor 9" cap screws.

The Valence batteries require fiberglass channel hold downs (McMaster Carr 8529K61). You will need four 5 ft. lengths of fiberglass channel for the full complement of sixteen batteries. The hold downs are secured to the bases with long cap screws and fender washers.

For Trojan batteries follow the above procedures without the provision for weld nuts. Following final welding up of the enclosure, weld in the 10 mm x 254 mm (3/8" x 10") L bolts (McMaster Carr #91587A120) which secure the hold downs. The center bolts will lie near the top of the angles and be welded from the top. The side bolts can be located lower down and be welded from top and bottom of the hook.

Trojan 1275 batteries are held down with specially fabricated steel straps and fender washers as shown in **Drawing 4.** The battery mounts and straps can be finished by painting or by galvanizing to resist rust. For lead-acid batteries because of the likelihood of acid spills and corrosive fumes, powder coating is a better alternative and provides a very attractive finish.

Securing the battery mounts in the truck follows the same procedure as for lithium-ion mounts described above.

Electronics Enclosure for Lithium-Ion Cells

Drawing 5 shows a box for individual lithium cells mounted as shown in **Drawing 3** with their BMS regulators mounted directly on the batteries. In this case, as mentioned earlier, it is necessary to enclose the batteries and their control electronics in the same protective environment as the power electronics, requiring a big, custom-made box.

The box has an open bottom to allow it to be installed after the batteries are mounted in the truck bed. It is mounted directly on the floor of the bed with ¼" sheet metal screws. The box is fabricated from aluminum sheet as shown. The rear surface is rebent to provide a 1" channel at the top for additional stiffness because this long panel suffers a lot of stress and vibration.

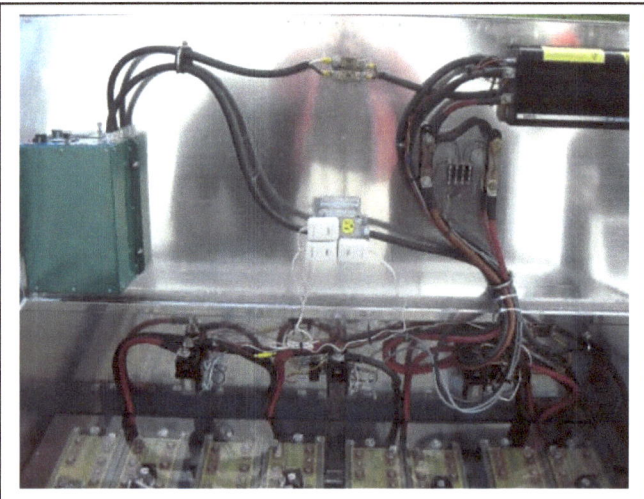

Figure 5.1 Electrical components installed

The top is shown plain (non diamond plate) because all the power electronic components are mounted to the top as shown in Figure 5.1. Each of the mounting screws must be sealed with RTV silicone rubber to prevent ingress of water and this is much easier on a flat top. You may also want to mount a heat sink on the top to keep the controller cool. Note that the top is 20 mm wider and longer than the box to provide ample overlap as the lid closes. The top is provided with a lifting handle and a hasp to secure it in place.

The right side of the box has a large hole lining up with the 2" conduit for the motor wiring to exit the box through the flat side of the truck bed. All four sides should be welded up using MIG or TIG welding. Note that the front surface has a notch for the hinged top. If you are less concerned with appearances this can be eliminated and a plain front to top joint can be used.

Since the batteries are in the same compartment as the power electronics, there is no need for Anderson connectors. The batteries are wired directly to three contactors which are mounted on the front face of the box with ¼-20 aluminum rivet nuts (McMaster Carr 94020A343). A fourth set of rivet nuts is used to mount the contactor between the controller and the motors. All of the power electronic components are mounted on the underside of the top as shown above. The motor, control and instrument wires come back down through a massive cable bundle and are fed down through a pipe union and elbow to the 42" length of 2" conduit leading to the motors below.

Electronic Enclosure Cross Bed Tool Box for Valence Batteries

Drawing 6 shows a typical layout for the power electronics for lead-acid or Valence batteries in a cross bed tool box. The box needs to be modified to permit connection to the batteries and the motor. A convenient way to connect to the batteries is to use a 350 Amp-rated Anderson connector for each group of four batteries, as shown in the sketch at the bottom of **Drawing 6**. The lower front wall of the box is prepared by using a 1 3/4" hole saw and a saber saw to cut four oblong openings as shown in the upper left of **Drawing 6**. The connectors can be screwed directly to the bottom of the box but a more elegant installation is to center the connectors in the holes by the use of 16 ga. brackets as shown in the detail. The connectors allow the box to be disconnected from the batteries and slid rearwards by two feet. This provides access to the batteries which remain solidly mounted in place in the truck bed.

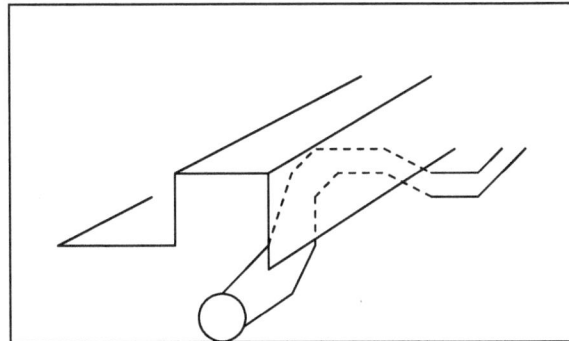

The elbow for the 2" conduit to the motor (McMaster Carr 7119K96) is mounted in the center of the box. The spacer between the control box and the front of the truck bed is shown in **Drawing 6.** It consists of a sheet of 1/16" aluminum diamond plate bent into an inverted gutter to fill the space between the tool box and the front of the truck bed while allowing the 2" conduit for the motor wires to rise up as the tool box is shoved forward over the batteries as shown here.

The electrical components all mount on the bottom of the box as shown in the lower left sketch in **Drawing 6.** Hole dimensions are given for mounting Albright contactors, a Curtis 1231 controller and a Zivan NG-3 charger. The routing for the high current wiring is shown. Since the controller functions as a current transformer #1 or 1/0 wiring is adequate for the connections to the batteries, but 2/0 wiring is preferred for connection to the motor.

The AC inlet receptacle is shown in a sketch on the right of **Drawing 6.** Make 1 of 16 ga aluminum and mount as shown. The inlet itself should be a waterproof male receptacle (McMaster Carr 69435K262 to mate with female plug 69435K243). Alternatively the SAE J-1722 connector pair specifically designed for electric vehicles can be used. This is more expensive, but is recognized and compatible with public charging stations which are becoming available and will grow in the future. In either case it is much more convenient to have a receptacle at the vehicle end rather than having a long cord on the vehicle to connect to the AC power source. Mounting the receptacle on the rear surface of the box runs into the same problem with the knuckle as the 2" conduit on the front. To resolve this mount the receptacle high up on the rear of the box at a convenient distance from the edge of the truck bed. The AC wiring inside the box should be 3-wire #10 rubber-insulated cord suitable for 30 amp 240 V service.

The same 3-wire #10 cord should be used for the extension cord to run from the AC source to the vehicle. Provide enough cord to connect to the vehicle comfortably without crossing it or getting under the wheels. Ideally the source of AC should be on the same side of the vehicle as the charging receptacle, and the entire installation should be mounted in a garage providing protection from the weather.

Charging stations specifically designed for electric vehicles are available from a number of manufacturers. They are expensive, but subsidies may be available to offset some of the cost. Installation should be by a licensed electrician, although in New Jersey such an installation has been declared to be "Minor Work" which can be done by anyone, as long as it is permitted and inspected. "Smart grid" installations are being developed which will eventually allow for time-of-use pricing. A handy accessory for a home installation is a timer which will allow you to plug your vehicle in when you get home but hold off charging it until the batteries have cooled off and the power may be cheaper. A separate kilowatt hour meter to keep track of the vehicle usage is also a useful accessory.

There is room in the cross bed box for tools and other equipment. To utilize this room make a shelf out of 1/8" aluminum as shown with a notch to provide for the AC input wire and a hole over the Zivan charger to give visual access to the LED display on its top surface. The ends of the shelf are turned up to provide for airflow into the electronic compartment for cooling, and provide a grip to raise the shelf. A better alternative for following the charger is the Zivan temperature probe which should be passed down to the batteries through one of the Anderson holes and which has a separate lead with an LED indicator at the end which can be passed out through a 3mm hole in the rear of the box to show charger status without opening the top.

The Poly Chain Drive: Overview

Since the drives for the Big Twin installation with Agni motors and with 9" series wound motors are very similar, they will be described together. The only difference is that the series wound motors need smooth faced reverse rollers and the motors are mounted 51 mm below the drive shaft centerline to clear the underside of the truck which is not necessary with Agni motors. One drive may be converted into the other by redrilling the motor mounting brackets and adding the weld nuts to mount the rollers. The series wound motors also need a support at the rear end and the electrical boxes are different. **Drawing 8** shows an elevation view of the active components of the Poly Chain drive for twin Agni motors and **Drawing 13** shows the same view for twin 9" series wound motors, both in cross section at the location of the Poly Chain sprockets. The components are viewed from the front and shown relative to the frame rails and the exhaust system. Purchased and specially made part numbers are keyed to the parts list, **Drawing 30**. The idler assemblies are symmetrical around the center of the drive shaft, which is offset by 33 mm to the right of the truck frame. A dimensioned drawing of the truck itself at this location is shown in **Drawing 18.** The left side of **Drawing 13** shows how the 1200 mm Poly Chain is installed without the reverse idler. The right side shows the reverse idler installed and the idlers separated to tension the Poly Chain.

Drawings 9 and 14 show plan views of the complete drive mounted in the truck looking down, along with the dimensions of the truck itself. The motors are offset axially by 31 mm to align with the two driven sprockets. The left motor is further to the rear to clear the fuel lines leading forward from the fuel filter. There is a distance piece on the drive shaft setting the sprockets 22 mm to the rear for the same reason.

Drawings 10 and 15 show side elevations of the compete drive mounted in the truck, and in **Drawing 10** a dimensioned drawing of the truck itself. The drive line slopes down 4.5 degrees to the rear, and the electric motor mounts are tilted to match this and keep the Poly Chain runs straight. The front end of the motors is supported by formed steel brackets and a formed steel cross piece shown in detail in **Drawings 11 and 16**. The rear of the 9" motors are supported by a 2" channel secured to the top of the frame rails by 12" J bolts and with steel straps to the tapped holes on the aluminum motor end bells as shown in **Drawing 26**.

There are four major components of the base structure. The upper part of **Drawings 11and 16** show the 3/16" steel plate brackets which secure the individual motors, idlers and support posts. The lower part of the drawing shows the formed 3/16" steel support channel and the assembly of the entire structure from the top. The right side shows the 3/16" steel clips that attach the support structure to the truck frame rails without interfering with the integrity of the frame.

Fabrication of the Twin Motor Drive

Unless you have a fully equipped shop, it would be wise to farm out fabrication of the 3/16" steel parts to a sheet metal fabricator who can handle tight bends in heavy material. The motor brackets are fairly straight-forward to lay out and drill. Bending them is a challenge. They could be fabricated by welding, but be sure that they are strong. The 9"motors weigh 170 lbs each, and at some point during assembly they will be twisting those brackets.

The Drive Base

The formed support channel is designed to be a stiff, strong support with absolute minimum depth which detracts from the ground clearance of the truck. They could be fabricated from angle or channel iron but with some sacrifice in utility.

The brackets are welded to the channel as shown. Note the set back of the left bracket. Make the brackets accurately parallel to keep the Poly Chains aligned. The assembly is located on the frame by the right bracket and the stop welded to the outside of the support channel.

The clips shown at the right of **Drawings 11 and 16** secure the right bracket to the top of the right rail with two 3/8-24 grade 5 screws and lock washers, and the left side to the bottom of the left rail similarly.

Shafting & Driven Sprocket

Drawings 12 and 17 show the modifications to the drive shaft to mount the driven sprockets, the driven sprocket assembly, the idler brackets, balance springs and support posts, and the electric connection boxes mounted on the rear of the motors. **Drawing 17** includes details of the smooth-faced reverse rollers.

The driven sprockets need to be bored out to 127 mm to clear the vibration damper ring and the front universal joint on the drive shaft. First fabricate the 1 1/4" thick aluminum distance piece on a face plate. You may want to turn the 3 mm relief in the front face of the distance piece to clear the nose of the transmission boot before doing anything else. It does not have to be exactly concentric with the rest of the assembly. Turn the piece around and bolt it to the face plate on a piece of plywood so you can finish the OD at 169 mm or 180 mm right to the edge. You can now do all the rest of the machining in one set up to get an accurate concentricity between the 1.598" hole to receive the front yoke of the universal joint and the nominal 155.95 mm OD to fit the inner diameter of the 67 tooth sprocket for the Agni or the 183.64 mm ID of the 80 tooth sprocket for the 9" motors. This diameter can vary from sprocket to sprocket, so measure it accurately and turn the aluminum to fit.

Bore the center to 1.598" +0.001-0.000. The front yoke is nominally 1.598" in diameter, but there is likely to be a little rust on it, which makes the fit tight. If in doubt measure it. Chamfer the inside edge of the center hole to clear the radius on the yoke and allow the distance piece to lie flat on the vibration damper ring on the yoke. Turn a step in the rear 10 mm of the OD to match the sprocket ID ±0.001". Scribe bolt circles with two concentric hole patterns at 79.0 and 140.0 mm for bolting to the drive shaft and bolting the distance piece to the sprocket as shown with a pointed tool while the distance piece is still in the lathe. In this way all the critical dimensions are completed in the same setup and are guaranteed concentric and perpendicular.

Drill the distance piece. The outer 12 hole pattern secures the sprockets with alternating 5/16" heat treated screws and 5/16" roll pins. The inner 6 hole pattern secures the plate to the vibration damper ring on the front yoke of the drive shaft with four 5/16" grade 5 screws and two 5/16" dowel pins.

Center the 62 mm thick sprocket in a lathe relative to the tapered center hole which should be concentric with the ID of the sprocket flange that mates with the distance piece. Bore out the center to 127 mm to clear the vibration damper. Assemble the distance piece and sprocket and drill through the twelve hole pattern, installing screws as you go to hold the alignment. Ream the alternate holes for the roll pins and match mark all the parts to ensure correct assembly in the future. The sprocket assembly is installed on the drive shaft as described on Chapter 3. The only modification to the drive shaft assembly is to drill and tap the screw holes and to drill and ream the dowel pin mounting holes with the sprocket assembly clamped in place.

Idler Brackets

Drawings 12 and 17 show the idler support assemblies for the idlers on the front of the drive. The idler rollers themselves are purchased parts with 32 tooth sprockets mounted on ball bearings. A total of four are required. They are supplied with mounting bolts having a ¾-16 thread. In order to reduce the mass of the idlers and improve their dynamic response, as well as to make a compact installation, the bolts are modified as shown in the upper left corner of **Drawings 12 and 17**. The massive bolt head is thinned down to 4 mm and the shank of the bolt is drilled out to 14 mm (9/16"). The length of the bolt is reduced to 73mm.

The OD of the bolts is 20 mm to match the bore of the idler ball bearings. Four pairs of 1/8" thick steel idler brackets are needed, each pair having one hole at 20.1mm and one at ¾" (19.1mm) to provide equal support to the two ends of the mounting bolts. The other end of the brackets is reamed to press fit onto a 7/8" OD x 0.065" wall 4130 steel tube. A Critchley adjustable reamer is convenient for this. Otherwise a standard reamer may be used and the tube accurately ground down to match.

The idler brackets are assembled onto the 4130 tubes to create pairs of upper and lower brackets as shown. The idler bracket plates are assembled onto the tubes with the smaller 0.750" (19.1 mm) holes toward the bottom (lever) end so that the mounting bolt heads are up and the ¾-16 jam nuts and lock washers securing the idler sprockets are down (nearest the motor mounting brackets and the motors). This allows the bolts to be withdrawn and the idler sprockets to be removed from the idler brackets without removing the brackets. It is highly desirable to make dummy idlers to hold the spacing of these plates while they are being welded to the tubes. The dummies should allow for ¾" SAE washers on each side of the idler sprockets for a total spacing of 53 mm.

Once the assemblies are stacked up in pairs they may be finished by welding or brazing the plate members to the tubes. TIG welding is preferred for small lots. Silver soldering is not recommended on strength grounds. These components are highly stressed in service and must not fail.

Following welding, the parts should be galvanized or painted for corrosion protection and appearance. The idler sprockets are assembled into the brackets as shown with the sprockets oriented inwards toward the motors, secured by the modified bolts and with a ¾" SAE washer on each side for spacing. The bolts are secured with 3/4"-16 jam nuts and ¾"internal tooth lock washers.

The 3/4-16 jam nuts for the lower idler brackets are notched for the 3/32" tensioning cable as shown. If desired the idler sprockets can be lightened for better dynamic response by boring out the center of a 32 tooth sprocket and making a new aluminum bearing assembly as shown in **Drawing 12**.

The idler brackets are mounted on 100 mm long cold rolled steel posts tuned to fit the bore of the 7/8" tubes with clearance for free rotation. The posts are welded into ¾" (C size) holes in the motor support brackets

The final parts for the idler assemblies are the support plates shown below the idler support assemblies in **Drawings 12 and 17**. These tie the tops of the idler posts together and support them from the base plate to ensure that the idler support posts do not bend under the stress of operation. The top of the support is a triangular plate of 1/8" steel with ¾" holes to match the post locations. This is welded to a 5/8" OD x 0.065" wall 4130 tube spacer and a ½" washer as shown. The tube is secured to the base plate with a 12 mm x 100 mm hex head cap screw and a nylon lock nut.

The balance springs to support the weight of the idlers and equalize the forces on the driven sprockets are shown in **Drawings 12 and 17**. They should be wound on a 1.25" diameter mandrel to give 1 ¾ turns, one left and one right with end hooks as shown. You may need to heat the ends to form the hooks without cracking the wire.

Drawing 12 shows details of the 7" square plastic junction box for the right hand Agni motor to contain all of the electrical connections for power, control and instrumentation. A plastic laminate plate to insulate the left hand motor brush assembly is also shown.

Drawing 17 shows details of the smooth faced idler rollers made from 36 tooth 8 mm pitch idler sprockets with the teeth turned off and the mounting bolts modified to shorten them and add holes for the locking wire. There is also a mounting spacer to be welded to the motor support brackets for these rollers. **Drawing 17** shows the electrical junction boxes that cover the brush ends of the motors.

The Single Motor Drive

Drawing 18 shows an elevation view of the active components of the Poly Chain drive for a single motor. The components are mounted on a steel plate and are viewed from both sides relative to the bottom of the frame rails at the plate center line. It also shows a dimensioned drawing of the truck where the main sprockets are mounted on the drive shaft in cross section. Note that the two sets of bearings are not at the same height. This is because it is necessary to drop the motor 40 mm relative to the centerline of the drive shaft to clear the structure of the truck with 11" motors. The counter shaft side is correspondingly raised 40 mm to keep all the Poly Chain runs symmetrical. The idler assemblies are symmetrical around the center of the drive shaft.

Drawing 19 shows a plan view of the compete single motor drive mounted in the truck looking down with the outlines of Warp 11 and AC Propulsion motors. The plan view shows the two sets of idlers on the drive side lining up with the two sprockets on the drive shaft and tensioning the drive chains. There are two idlers on the motor side tensioning the cross chain which drives the counter shaft from the single motor through a differential tensioner which maintains equal torque on the motor shaft and the counter shaft.

Drawing 20 shows an elevation of the complete drive mounted in the truck from the left side, along with the dimensions of the truck itself and outlines of the two applicable motors. Purchased and specially made parts are number keyed to the parts list in **Drawing 30**.

The base of the drive is a welded sheet metal structure shown in **Drawing 21.** There are four major components of the base. At the upper left of **Drawing 21** is the 12 ga. steel plate with weld nuts which secure the idlers and tensioning cable pulleys.

At the upper right are flat and bent up drawings of the 12 ga. rear plates which are edge welded to the front plate to form the vertical structure that holds the drive components in alignment. These plates also have weld nuts that secure the components on the rear (motor) side of the drive.

At the lower left are two views of the support channel bent up out of 3/16" steel which secures the drive to the frame rails of the truck. There are also views of the vertical structure welded to the support channel showing the orientation of the parts.
Finally, on the lower right there is a sketch of two mounting clips of 3/16 steel that secure the entire assembly to the frame rails.

Fabrication of the Drive

To begin fabrication, cut out the front plate as shown in **Drawing 21** and drill all the holes for the weld nuts. Note that the ¾ -16 nuts are actually jam nuts that take a ¾ "hole and are located with a ¾" screw for welding. The ½-20 and 3/8-16 nuts are available as weld nuts with projections per the parts list, **Drawing 30**, and require oversize holes as shown. Weld in all the nuts, being sure to weld them on the correct side. The drawing shows the plate from the rear side where the nuts are.

Next cut out the rear plates as shown in the flat, noting the vertical offset between the hole patterns. Bend up the rear plates and weld in the weld nuts as shown. In this case the nuts are on the <u>other</u> side. Both front and rear plates are shown from the rear looking forward.

Note that the right hand rear plate has four weld nuts to secure the motor mount spacer shown in **Drawing 25.** This component should be fabricated next out of 3/16" steel. It is a box with a top and right side to rigidly connect the drive to the electric motor while providing clearance for the cross chain between the motor and counter shafts and for its installation through the open bottom and left side. The square cut out mates with the right hand plate when the spacer is bolted to the rear of the drive. The round hole in the opposite side centers the electric motor. Drilling patterns for Warp 11" and AC Propulsion AC-150 motors are shown. The hole pattern for the Mayr 6500 electric clutch magnet is also shown. The top of the spacer box needs to be bumped up as shown to provide adequate clearance for the cross chain.

Assembly of the drive requires alignment jigs shown in **Drawing 24.** The longer jig simulates the electric motor with a very long rigid shaft. The shorter one is to align the counter shaft bearings. Weld the shafts to the plates and stress relieve as necessary. Then turn the surface of the plates to be accurately perpendicular to the shafts.

Assembly of the drive begins with bolting the long jig to the motor mount spacer box with the shaft sticking through the square opening on the opposite side of the box. The rear plate with a 1" mounted ball bearing is screwed to the spacer box and 15/64 " holes are drilled and reamed to ¼"through the plate and the spacer to dowel the two together for exact alignment.

Another mounted ball bearing is now fastened to the forward drive plate, which is mated to the rear plate assembly. At this point the motor mount, the rear plate and the front plate should be accurately aligned by the jig and the rear plate is now welded to the front plate along the edges of the flange in the rear plate. The front and rear plates should be aligned along the top leaving the bottom of the rear late 3 mm higher than the front. This will provide the proper 4.5 degree angle of tilt when the plates are welded to the support channel.

The lower, short jig aligns the counter shaft. The two holes match the ½-20 mounting holes for the bearing on the rear plate. Bolt the jig to the plate, then install a bearing on the front plate and assemble the rear plate to it with the rod projecting through the bearing. The two bearing mounts are now aligned and the rear plate may be welded to the front similarly to the motor side.

The support channel is bent up out of 3/16" steel as shown and drilled for bolting to the mounting clips. The vertical assembly is welded to the channel as shown with the ribs of the channel projecting downward to stiffen it and the flat surface on top to interface with the bottom of the frame rails. 12 ga. stops are welded to the support channel as shown to position it laterally relative to the rails.

Finally, the mounting clips are fabricated from 3/16" steel and drilled to match the support channel. A much more finished appearance as well as corrosion resistance can be achieved by galvanizing all of these parts after welding.

Idler Brackets

Drawing 22 shows the idler bracket assemblies supporting the idlers on front and rear of the drive. A total of six are required. They are supplied with mounting bolts which should be modified as shown in the upper left corner of the drawing.

Six pairs of 1/8" thick steel idler brackets are needed, each pair having one hole at 20.1mm and one at ¾" (19.1mm) to provide equal support to the two ends of the bearing mounting bolts. The other end of the brackets is reamed to press fit onto a 7/8" OD x 0.065" wall 4130 steel tube as described for the twin motor drive. For the four forward idlers short tensioning levers of 1/8" steel are needed with holes similarly reamed. The interior of the 7/8" tubes must be reamed to ¾" diameter to fit over ¾" diameter mounting studs. Alternatively, the mounting studs can be turned down by 0.005", or better yet, special unthreaded studs can be fabricated and welded to the backing plates as described for the twin drives.

The idler brackets are assembled onto the 4130 tubes to create pairs of long and short brackets to match the right (motor) side and left (counter) side of the drive. The idler bracket plates are assembled onto the tubes with the smaller ¾" holes toward the bottom (lever) end so that the narrow mounting bolt heads are up and the ¾-16 jam nuts securing the idler sprockets are down. It is highly desirable to make dummy idlers to hold the spacing of these plates while they are being welded to the tubes. The dummies should allow for ¾" SAE washers on each side of the idler sprockets for a total spacing of 53 mm. The actuating levers are pressed onto the bottom of the tube assemblies as shown. Each pair of short and long assemblies consists of a left handed and a right handed assembly as shown in **Drawing 22**. Once the assemblies are stacked up in pairs, they may be finished by welding or brazing the plate members to the tubes. Silver solder is not recommended by reason of strength and reliability.

The two assemblies for the rear side of the drive are similar except that the levers are longer and may include overrunning clutches for self adjustment as shown in the lower left center of **Drawing 22**. The levers are fabricated of 1/8" steel as before. Stack the parts as shown in the center of the drawing, noting that there are two different angles for the lever arms, as shown. Weld or braze up the assemblies as with the front brackets

Following welding the parts should be galvanized or painted for corrosion protection and appearance. The idler sprockets are assembled into the brackets as shown with the sprockets toward the top secured by the modified bolts and with a ¾" SAE washer on each side for spacing. The bolts are secured with 3/4"-16 jam nuts and ¾"internal tooth lock washers.

For self adjustment with CSK-25 overrunning clutches, the 7/8" tubes of the rear idler brackets are shortened by 15 mm and a 15mm piece of 2 3/8" OD tubing is welded to the bottom of the lever and bored out to a press fit to the external 52 mm diameter of the clutch. The lever is cranked 15 mm to bring the tensioning end down to the proper height as shown. A ¾-16 x 1" high hex nut is turned down to a press fit to the 25 mm bore of the clutch to provide a base for the assembly. Galvanize these parts before assembly as it cannot be done afterwards.

Also before final assembly it is important to ream the interior of the 7/8" tubes to ¾" because it cannot be done once the clutch is installed. Again the better solution is a specially turned stud. With all the parts finished and welded, press the nut into the bore of the clutch being sure that the clutches will allow the levers to be pulled to the left looking down on the assemblies from the top as shown in the drawing. Then press the clutch assemblies into the bracket assemblies to complete the job.

A simpler alternative for self adjusting is shown in the lower right center of **Drawing 22**. The levers in this case are straight with a bent down tab at the end which engages a short piece of 3/8" square rack mounted on the backing plate to allow take up in one direction only.

The final parts for the idler brackets are the supports shown on the right of **Drawing 22**. These tie the tops of the front side idlers together and support them from the base plate to ensure that the ¾" support bolts do not bend under the stress of operation. They are fabricated as described above for the twin motor drive.

Shafting and Differential Tensioner

Drawing 24 shows the 1" keyed shafts which are purchased in 18" lengths and cut down as shown at the top. It also shows the modified drive shaft for the Mayr electric clutch.

Drawing 23 shows details of the differential tensioner, the Mayr electric clutch and couplings for single motors and the drive shaft. At the top left are shown the purchased high speed mounted ball bearings used to support the shafts. The bore is 1" (25.4 mm), and the bearings have a collet arrangement to center and secure the shaft.

Immediately below the bearings are shown drawings for the differential tensioner, a planetary geared coupling designed to provide torsional compliance to make up for residual misalignment between the front and rear Poly Chain assemblies and to reduce noise and vibration This component has not been built and tested, but all the other approaches to maintaining synchronism between the two sides of a single motor drive have failed, as described in Chapter 1. This one is at least theoretically sound. The differential gear set ensures that equal torque is applied to the right and left driving sprockets no matter how much the individual Poly Chains may stretch or wear.

The design has been worked out with standard parts from Boston Gear. The right hand gear and the pinions are stock parts, unmodified. The left hand gear needs to be bored out to 35.0 mm and rebroached for the 1/4" key. The key needs to be permanently fixed to the gear by press fitting, welding or locktite.

A 133 mm diameter steel housing is made to fit the motor shaft. As drawn it will fit any C face motor with a 1 1/8" shaft such as the Warp 9 and 11. An adapter for the AC Propulsion motor is shown below.

The steel cover of the tensioner carries a bronze oilite bearing that fits the OD of the modified gear which is keyed to a steel coupling sleeve that floats on the 1" driven shaft. The driven shaft passes through the drive to the right hand drive sprocket. The coupling sleeve is also keyed to the Taperlok bushing of a 40 tooth sprocket which drives the left hand drive sprocket through the cross chain and counter shaft.

Machining the cover and the body of the tensioner should be straightforward lathe jobs. The cover is a simple 7/8" thick steel plate bored out to fit the shortened bronze bushing and faced on the inside for the needle thrust bearing. It could be made of aluminum with a steel thrust washer for the needle bearing to run on.

The body is turned out of a 6" diameter rod, again steel or aluminum with a steel washer. With aluminum one might want to go to eight or twelve fastening screws around the periphery since there is not a lot of meat there. Note the flats that have to be milled into the bore to provide flat surfaces for the pinions to run on. Note also that the motor end is slit two ways, half way down perpendicular to the shaft and halfway through parallel to the shaft to allow the 5/16"-24 clamping bolt to tighten it on to the motor shaft.

To assemble, mount the body on the motor shaft and add the right hand needle thrust bearing. Insert the right hand gear, which has a ¼" key for the driven shaft permanently installed in its bore with Locktite. Install the three planetary pinions with the specially made 3/4 -16 cap screws and add the permanently keyed left gear with its coupling sleeve and the left hand needle thrust bearing. Fasten the cover with six 5/16" socket head screws to complete the differential unit.

The motor with the differential tensioner is now mounted on the spacer box of **Drawing 25** and the cross belt is looped around the 40 tooth sprocket. The box is mounted on the drive and the driven shaft is inserted from the front through the front bearing the cross belt and the differential tensioner until it engages the right gear with its permanent key. Lock the driven shaft to the front bearing with its collet. The remaining installation to line up with the driven sprockets and lock the driving sprockets is described in Chapter 3. The entire drive should be assembled out of the vehicle and prealigned while mounted on a wooden cradle. It is almost impossible to do this once the drive is in the vehicle. Cross chain replacement is done by loosening the front bearing and the clamp bolt on the differential tensioner and sliding the bearing, driven shaft and tensioner rearwards to expose a gap between the tensioner and the motor shaft.

Clutches

Theoretically, there is a good case to be made to include a clutch to disconnect the electric drive from the drive shaft, especially on the single motor drive. The clutch will avoid windmilling the electric motor when operating on gasoline, and it eliminates the speed restriction imposed by the rpm limit on the electric motor (not a problem with the AC Propulsion motor). Since the nominal speed limit on the Warp 11 motor is 5000 RPM, the top speed of the vehicle is limited to less than 60 MPH to avoid damaging the motor.

The most promising candidate is an electric clutch supplied by Mayr denominated 6/500.3010/24/45/40/3.This is a "double flow" design featuring high torque and high speed (thus high power) capability in a very compact and economical package. The size 6 is capable of 160 Nm at 5800 RPM or 97 kW transferred on 38 Watts of 24 V power for the magnet. The total weight of the clutch is 4.46 kg with the output hub. The clutch consists of three parts the magnet, which is stationary; the input rotor, which mounts on the motor shaft; and the output clutch plate and hub assembly, which mounts on the driven shaft.

Figure 5.2 Mayr clutch installed

Drawing 23 provides details of mounting the Mayr clutch that will fit the Warp11 inch motor. The magnet is shown mounted to the motor support box with six M6 screws and 15 mm spacers. An adapter is needed to go from the 1/18" motor shaft to the 45 mm bore of the input rotor. This is machined from 57 mm (2 ¼") steel bar stock as shown. The end of the adapter is machined to take a push fit 17 mm bore ball bearing for alignment. It is counter sunk for a 5/16" 18 flat head screw to secure it to the end of the motor shaft and threaded 7/16-20 for a jacking screw to separate it from the motor shaft on disassembly. The outside is ground to an ANSI standard interference fit with the 45 mm bore and milled for a DIN 14 mm key. When complete, the adapter is pressed into the bore of the input rotor to make a single unit.

The output hub also requires an adapter to fit it to the 1" driven shaft. This is machined from 50 mm (2") rod as shown in **Drawing 23**. This also is press fit into the hub as a permanent part with a 12 mm DIN key with the keyway reduced to a 2.5 mm depth because of limited space between the bore and the shaft. The alignment ball bearing is press fit to the front of the adapter, and the adapter is located on the 1" shaft by two 1/4-20 set screws, one bearing on the ¼" square key and one drilled into the shaft itself. The complete assembly with the output plate mounted on the shaft and the input rotor with its adapter can be seen in Figures 3.15 and 5.2.

The clutch suffered from severe misalignment in the testing done with it. It appeared that it would probably have adequate torque capability if properly installed, but since there does not appear to be an appreciable energy penalty from windmilling, and the clutch represents a considerable complication and expense, it has not been followed further. It is available as an option for those for whom high speed is essential.

Main Sprocket and Drive Shaft Modifications for Single Motor Drive

Drawing 24 shows modifications to fit the driven sprocket to the drive shaft. They are applicable to sprockets of 71, 75 and 80 teeth, but only the 71 tooth sprocket is consistent with the shaft spacing and idler sprocket layout shown for the single motor drive.

Drawing 24 shows both the method shown in **Drawings 12 and 17** with a single 62 mm wide sprocket modified and assembled with a shortened distance piece and an earlier method based on two 21 mm wide sprockets. The two sprockets need to be bolted together and to a plate centering them on the drive shaft using the front yoke of the shaft as a pilot. The downside is more complexity in assembly. The advantage is that there is a flange in the middle of the sprocket positively separating the two driving chains.

To use two sprockets a ¼" thick steel plate is turned to a close fit to the inner diameter of the sprocket rim and drilled with two concentric hole patterns as shown. The outer 12 hole pattern secures the sprockets with alternating 5/16" heat treated screws and 5/16" roll pins. The inner 6 hole pattern secures the plate to the vibration dampening ring on the front yoke of the drive shaft with four 5/16" grade 5 screws and two 5/16" dowel pins. Drill the screw holes full size and the pin holes 19/64" to allow for reaming. Bore the center hole 1.598" to center the plate accurately on the front yoke.

Chuck and center one of the sprockets in a lathe and cut out the center to a diameter of 4.500 inches. This sprocket will now center on the hub of the other sprocket.

Assemble the ¼" plate and both sprockets and drill through the twelve hole pattern of the plate, installing screws as you go to hold the alignment. Ream the holes for the roll pins and match mark all the parts to ensure correct assembly in the future.

Disassemble the sprockets and bore both of them to 127 mm to clear the vibration damper, then reassemble with screws and roll pins. The sprocket assembly is installed on the drive shaft as described on Chapter 3. The only modification to the drive shaft assembly is to drill and ream the six mounting holes with the sprocket assembly clamped in place.

Sheet Metal-- Motor Electronics Enclosure

An 18 gauge galvanized sheet steel enclosure shown in **Drawing 26** is screwed to the rear of the 11" motor to shield the wiring and the brushes from the elements. The rear of the box is solid, and the front is split to provide for the positive 2/0 power cable and the connection to the field. A 61 mm hole is provided in the rear for entrance of all the wiring from the electronics box (along with a supply of clean cooling air) via 2" flexible conduit. Two 22 mm holes are provided near the bottom for the multi conductor cable to the PLC in the cab and the one to the transmission. Since these receptacles have nuts only on the rear surface, a small rectangular sheet metal bracket is used to mount the receptacles and is secured to the interior of the enclosure with a sheet metal screw. Two 8mm drain holes are provided in the enclosure to eliminate condensation. The enclosure is screwed to the rear of the motor using the tapped holes in the rear end bell.

These boxes can be fabricated from a single 18 ga. galvanized sheet, by laying them out and drilling the holes first. Then bend the flanges for attaching the front covers noting that the bottom flange is cut out to maximize ground clearance. Then bend up the sides and solder or weld them to form the box. Paint the welds with cold galvanizing paint for appearance and corrosion protection.

The heat shield shown in **Drawing 2** is required if you elect to leave the exhaust crossover pipe in place. It is secured to the crossover with two 2 1/2" muffler clamps. The square notch clears a clamp on the pipe and locates the heat shield transversely on it. The right side of the heat shield is bent away on a curve as shown to wrap around the pipe. It will be necessary to hammer a bash into the curved section with a ball peen hammer to fit around the main sprockets and idlers.

Rear Motor Support

Because the 11" motor weighs about 300 lb, it needs support independent of the drive mounting on the support channel. This may be accomplished with 1" x 3/32" steel straps screwed to a 2" channel member flush against the crossover and blocked up on the frame rails to clear the 11" motor electrical housing as shown in **Drawing 20.** Details are shown in **Drawing 26.**

Light Twin Drive

The light twin drive with 7" series-wound motors is shown in **Drawings 27-29.** Construction is essentially identical to that for the larger 9" motors.

Material Schedule

Drawing 30 is a purchased material schedule keyed to numbers on the drawings, and providing part numbers, primarily from the McMaster Carr catalog.

Chapter 6. Catalog of Purchased Components
and
Engineering Data

Component Suppliers

The following is a list of suppliers of motors, chargers, controllers and other components, many of whom we have used and found reliable.

AC Propulsion of San Dimas, CA. (http://www.acpropulsion.com) Tel (909) 592-5399 (Motors & controllers)

Advanced DC Motors of Syracuse , NY, (info@nidec.com) Tel. (315) 434-9303 (Series-wound DC motors)

Agni Motors (info@agnimotors.com) (Axial flux DC motors)

Curtis Instruments (curtisinstruments.com) (Controllers, accessories, instruments)

Electric Conversions, 515 North 10th St. Sacramento, CA, 95811, Tel. (916) 441-4161, (elcon@jps.net). (Chargers)

Evnetics, LLC, 2027 4th. Ave. S, St Petersburg, FL 33712, Tel. (727) 895-8989, (www.evnetics.com) (Controllers)

Gates Rubber Co., Denver, CO, (http://www.gates.com) Tel. (303) 744-19110 (Poly Chains)

High Performance Electric Vehicle Systems of Ontario, CA, (http://hpevs.com) Tel. (909) 923-1973 (AC Motors)

Manzanita Micro, 26125 Calvary Lane NE, Suite 300, Kingston, WA, 98346 (360) Tel. 297-1660 sales@manzanitamicro.com (Chargers, controllers, battery systems)

NetGain Motors, Inc. 800 S. State St., Suite 4, Lockport, IL 60411, Tel. (630) 243-9100, www.go-ev.com (Series-wound DC motors)

Trojan Battery Company, Tel. (800) 423-6569 (www.trojanbattery.com) (Lead-acid batteries)

Tucson EV at EV@TucsonEV.com (J-1772 charging equipment)

David Mosher of Cedar Rapids, IA, Tel. (319) 431-8094 (Curtis controller modifications)

Valence Technology, 12303 Technology Blvd., Suite 950, Austin, TX 78727, Tel. (512) 527-2900, or 1 (888) 825-3623, e-mail sales@valence.com. (Lithium-ion batteries)

Full line suppliers

EVA

Electric Vehicles of America P.O. Box 2037 Center St., Wolfeboro, NH 03894, Tel. (603) 569-2100, sales @ EVAmerica.com provide everything you need for electric vehicle conversions from batteries to instruments including motors and power electronics. I have dealt with them for twenty years and found them unfailingly helpful, fair and well informed.

KTA

KTA Services, Inc. 20330 Rancho Villa Rd., Ramona, CA, 92065 Tel (760)787-0896 (KTA-EV.com) are similar to EVA in the components and kits carried. I have not dealt with them myself, but they have been around awhile and have a substantial product line.

McMaster Carr Supply Co.

McMaster Carr (mcmaster.com) are nationwide. Their web site catalog gives you access to their 510,000 advertised products including every tool, fastening, bearing gear, chain etc, etc. ever invented, along with a complete selection of metals, plastics and other materials to create your project Their ubiquitous 3500 page yellow catalogs are a treasure trove of goodies and information. I have keyed components in the drawings to McMaster catalog numbers. They have local distributors all over the US, and if you are near one you will get next day service, and almost everything is in stock at reasonable prices.

Newark Electronics

Newark in Chicago, IL, (http://www.newark.com) is similar to McMaster Carr in the electronics field. They have a 2500 page catalog of semiconductors and all sorts of switches lamps resistors etc., etc. They were the source of the Crouzet PLC controller I used. I am not entirely enamored of the controller, but Newark are about as good as McMaster in terms of speed, percentage of items in stock (much improved since years ago) and prices.

Mouser Electronics of Texas at mouser.com Tel. (800) 346-6873 are similar to Newark in terms of inventory and service.

Digikey digikey.com Tel. (800) 344-4539 is another very popular electronics supplier.

Drive Components

Partnerships One, LLC, 32 Woodlane Rd., Lawrenceville, NJ 08648, Tel. (609) 896-2193, partnerships1@verizon.net can supply drive components per **Drawings 5, 11 and 12 or 16-17.** Also Agni motors and Valence batteries**.** Call or e-mail for quote. Since quantities are low, prices will be high, but competitive with other sources of the same quantities.

Batteries

The options for batteries are limited to those types which can provide heavy current draws for sustained periods. For lead-acid batteries this restricts the choices to flooded types such as golf cart batteries. Typically golf cart batteries are 6 or 8 V requiring 25-30 individual batteries to provide the required 200 V for acceptable performance. The total weight of almost a ton makes this option infeasible.

The use of heavy duty 12 V flooded batteries, specifically Trojan type T-1275s rated at 150 Amp hours at 20 hr (0.05C) and 120 Amp hrs at 0.2C, have proven marginally satisfactory. A minimum pack of twelve batteries weighs 984 lb. They can be grouped in groups of four for watering from a single point. Life can be expected to be about a year of continuous use. The price is approximately $200 per battery. Because of their limited life, this cost will be incurred approximately every year, which soon offsets the low initial cost, to say nothing of the inconvenience.

The simplest lithium-ion replacements are nominal 12V batteries from Valence Technology. They offer batteries in the popular Group 24 and group 27 sizes. The operating voltage is 12.8 V and the group 24 type U24-12XP is rated at 110 Amp hrs. These batteries are rated at 150 Amps continuous and 300 amps for 30 seconds, which is just adequate to propel a pickup truck. The battery weight is 15.8 kg each (34.76 lb), and the total sixteen battery pack for 205 V will weigh 556 lb, less than half the weight of a comparable lead-acid pack and leaving an ample payload for the vehicle.

The group 27 Valence U27-12XP is rated at 138 Ah with the same continuous and pulse ratings as the U24. It weighs 19.5 kg, 42.9 lb, or 687 lb for the sixteen battery pack. It is a better choice in terms of range and capacity per dollar.

Valence puts most of the intelligence into the batteries themselves. The pack is strung together with an RS-485 Can Bus data line and controlled by a single low-cost U-BMS-HV battery management system. The BMS monitors and controls charge voltage, discharge voltage, discharge current individual cell voltages and individual cell temperatures. It requires separate DC contactors for charge current and discharge current and can control separate precharge contactors for charging and discharging. It performs cell equalization as long as the pack is connected to the charger and the charger is providing a float voltage of 13.8 V per baattery. The performance of each cell can be viewed via a diagnostic kit attached to the RS-485 line at the opposite end from the BMS. Data on pack performance can be logged via a Can Bus connection to the BMS and a serial link to a laptop.

Valence batteries are the top of the line with prices to match. List price for the U24-12XP is $2000. For the U27-12XP it is $2500, although substantial discounts are available for full pack purchases. The BMS is very low in cost at $350, presumably to encourage its use. The diagnostic and Canbus kits are $300 each.

Chinese-made cells offer the low-cost lithium-ion approach. There are a number of vendors of whom Winston Battery (Thunder Sky), one of the first, is typical. They offer individual cells based on lithium iron phosphate cathode material, which is free from thermal runaway concerns, and claim excellent high current and long cycle life properties. The nominal voltage per cell is 3.2V and the Thunder Sky cells are offered in 40, 60 and 90 Amp-hour ratings. A 50 V pack requires sixteen cells in series and the total 200 V requires 64. The maximum continuous current is at the 3 C rate, so a single string of 90 Ah cells will provide up to 270 amps and 18.4 kWh of capacity. The cost will be approximately $7300 to which must be added the cost of the BMS. One such system is available costing $200 to monitor sixteen cells. The total is thus $8100. There are a number of packagers who will put together an integrated pack with BMS for about $10,000 per 20 kWh.

CALB (China Aeronautical Lithium Battery) is one of the best Chinese suppliers. CALB offers 40, 60, 70, 100, 130, 180 and 400 AH sizes in their SE line with prices and performance similar to Thunder Sky batteries. Manzanita Micro carry CALB cells and offer regulator boards to control 8 cell strings. The regulators connect to Manzanita chargers to prevent overcharging and to a display showing the condition of each cell via six wire telephone cable.

Manzanita also offer "Reg decks" which are large printed circuits with heavy copper bus bars attached to connect the eight cells in series and simultaneously make all the control and monitoring connections as well as having thermistors to monitor the temperature of each cell. These are very convenient in assembling a pack and present a neat appearance. They are somewhat expensive at close to $275 for the circuits and $320 for the reg. decks amounting to almost $60 extra per cell.

Manzanita has just introduced 12 V and 24 V 100 AH battery modules completely assembled with regulators for $1000 and $1800. This is a 72% premium over the cost of the cells, but represents a considerable saving in assembly cost.

Electric Vehicles of America offer Flux Power cells at 60, 90 100, 160 200, and 260 AH sizes and battery management systems at competitive prices

Chargers

Zivan chargers are available from the US distributor Electric Conversions and a number of vendors such as EVA. These are switching power supplies available in 1, 3 and 5 kW capacity. The 1 kW NG-1 unit operating on 120 V input and approximately 9 amps is too small to charge a 20 kWhr pack conveniently. The NG-3 operating on 240 V single phase and 14 amps is quite adequate. It can be programmed to any desired charging profile by Electric Conversions.

Manzanita Micro has supplied chargers for a number of years. Their units are characterized by maximum current capacity and they offer 20, 30, 40, 50 and 75 Amp units, all with 240 V and 120 V, single phase input. They are convenient in that the maximum current and voltage can be set manually. The output DC voltage can be varied from 12 to 450 V by a 20 turn trim pot and now in newer models by a much more convenient digital input switch. The current can be varied from 0 to maximum by a front panel knob. For J-1772 installations there is a rear panel electrical connector to a relay to switch current control to remote and for a 0-5V control signal. Because Manzanita chargers have no isolation, the ground and signal wires are hot and require optical isolation from the signal source. The chargers will automatically charge at constant (set) current to the selected voltage and then at constant voltage. There is a timer which is also manually settable for time at constant voltage after which the charger shuts down. There is an interface with the Manzanita regulators on the cells which will lower the charge current if any of the cells becomes overcharged to prevent cell damage.

There are a number of other charger manufacturers and suppliers. Metric Mind (metricmind.com) supplies high end, very expensive Brusa Chargers.

Home Charging Stations

Single phase 240 V power is universally available direct from residential service entries. It is used for electric ranges and clothes dryers. An additional 30 Amp circuit can readily be added by an electrician. There is now a standard plug for charging electric vehicles built to the SAE- 1772 standard. It is expensive and the NEMA 10-30, 6-50 or 10-50 three prong plugs and receptacles are perfectly adequate. The four prong grounded 14-30 and 14-50 are even better. The weather tight L6-30 or L14-30 twist lock pairs are good on the vehicle. The only problem is that you won't be able to charge at public charging stations.

If you want to get compatible with the SAE J-1772 standard, which is a very wise thing to do, based on safety and compatibility with the emerging public charging network, you can buy a J1772 charging cord and a receptacle from suppliers such as Tucson EV. They are more expensive that NEMA equivalents, but are coming down in price and offer advantages in eliminating the possibility of driving off while plugged in and other features that you can implement relatively easily if you understand the standard. The easiest solution is to buy a packaged charging station with the brains built in that just needs to be wired for 240 V AC. The most accessible supplier of this Electric Vehicle Service Equipment (EVSE) is Clipper Creek (clippercreek.com) who offer the LCS-25 charger for home installation. It can deliver 20 A of 240V AC (4.8 kW, level 2) and costs $495. They also supply a level 2 CS 40 charger and one specifically for the Tesla. There are federally funded programs to install free chargers in some locations which are worth looking into.

Coulomb Technologies (http://coulombtech.com) is perhaps the largest supplier but seem to be concentrating on their ChargePoint network of public chargers rather than supplying the home market. The network contains 10000 charging stations with an accompanying map. ChargePoint have an arrangement with Ford to supply 5000 free charging stations to Ford EV owners.

Eaton (http://www.eaton.com)offer the Pow-R-Station family of charging stations.

Ecotality was another maker of charging stations, trade named Blink, with a network for billing and 5000 stations. Their home charging offering was a level 2 wall mount. They also had a federally funded program to install a number of subsidized stations. They suffered from a reputation for unreliability, Their stations were often on the blink, and they have since gone out of business.

General Electric (http://www.geindustrial.com) offers the Wattstation.

Leviton evr-green (http://www.leviton.com)offer home charging stations with a prewire kit allowing for plugging in to a NEMA 6-50 240 V receptacle.

AeroVironment (avinc.com) offer a 240 V Level 2 home charger, and fast chargers with Chademo connectors. The EVSE-RS plugs in to an existing 240 V AC outlet and is portable, allowing the owner to take it to any premises so equipped. The unit with the required electronics and a long J1772 cord retails for $1099. [xx]

EV Collective (www.evcollective.com) offer battery buffer quick chargers with a large storage battery to accumulate energy from a relatively low wattage AC connection and deliver it through a high wattage DC connection to charge the vehicle in 30 minutes.

Nissan is offering a quick charger to match the CHAdeMO inlet now standard on the Leaf. [xxi] It will recharge the 24 kWh Leaf battery to 80 % in 30 minutes for a mere $15,500. Nisan has also developed a 6.6 kW V2H charger/inverter which offers emergency power to the home for $6000. It is currently available only in Japan.

There are a bunch of other EVSE sources and an open source design for homebuilders at http://code.google.com/p/open-evse/.

Controllers

The choice of PWM controllers for DC voltages above 144V has been very limited until recently. The only option has been the Zilla high voltage liquid-cooled unit from Café electronics.

These were very sophisticated units with 1000 and 2000 Amp capability and programming and data logging capabilities via an RS-232 port on the "Hairball" interface unit. After a difficult transition period, Zilla chargers are now supplied and supported by Manzanita Micro.

A more recent option is to modify the Curtis 1231-8600 controller to increase its voltage from 144 to 200 V and its current capability from 500 Amps (225 A, 1 hour) to 1000 A, 500 A continuous, still air-cooled. The conversions are done by David Mosher. The 1231 costs $1500 and the conversion will add around 50% to that. The Curtis pot box-throttle assembly with either of two styles of foot pedal will add another $100.

Netgain now offers high voltage controllers. The Netgain 160/1000 as the name implies can handle 160 V and 1000 amps (300 continuous, water cooled) and matches their Warp 9" motor. The price is $1900 plus $150 for an input device to provide a 0-5 V signal from a throttle input. The Netgain 360/1400 provides up to 360 V and 1400 A (still 300 A continuous) matching their Warp 11" motor and costing $4100.

Finally Evnetics (evnetics.com) now offer the Soliton 1 capable of 340 V and 1000 A, continuous, water cooled for $2900. A Soliton Junior was released in 2011 offering 340 V, 450-500 Amps, continuous for $2100 plus $150 for the input device. This controller, available from EVA and other distributors, is the best match to our requirements. It has the high voltage capability needed, it is easily programmed through a USB link, and it offers enough current capability to power a truck while limiting the current automatically on hot days if overloaded.

For AC motors Curtis offers the 1238R-75xx at 72-96V and 550 A max, and the 1238R-76xx at 72-96 V and 650 A max. Whereas DC controllers can control current to two motors in series or parallel, AC controllers can control only one motor, although the Curtis controllers do have a feature which can synchronize two motors to the same speed. This makes the AC system somewhat more expensive than an equivalent DC system.

Wiring

The main DC circuit includes the battery interconnections as well as the wiring inside the control cabinet and the 9 ft run down through the bed of the vehicle to the motor below. This adds up to a surprising 50 feet of wire carrying up to 300 Amps. Number 1/0 wire will carry 85 maximum amps and will have a voltage drop of only 1.2 V at 220 amps average current. It is recommended that at least this gauge wire be used. Fine stranded welding cable is recommended with high quality crimp fittings for termination. Copper prices fluctuate widely but currently (Spring, 2014) 2/0 welding cable is close to $4.50 per ft and lugs are $3.00 each. You will need approximately 50 feet and 50 lugs or $375 worth of high current conductors.

It is imperative to cut this wire cleanly because the lugs fit it tightly and there is no margin for a rough end. A cable cutter will cost around $70 and save hours of frustration.

Crimps can be made with a low-cost hammer tool for about $40, but a more professional job requires a rotating die crimper for $250. The terminations can be finished and sealed with heat shrink tubing lined with hot melt adhesive and a heat gun. All of these tools may be rented from EVA for a modest fee. It pays to plan ahead and do it all at once.

Motors

Radial Field DC

For conventional radial field series-wound DC motors the options are Advanced DC motors and Netgain.

Advanced DC has been supplying medium hp traction motors for many years in sizes from 7 to 9" diameter and power ratings from 10-30 HP. Of particular interest is the 9" FB1-4001A motor rated at 120 V but operable to 144 V. At 220 Amps it will deliver 36 ft lb of torque and 36 HP at 5200 RPM, which is its maximum speed. This motor is marginal for heavy vehicles because of its restricted voltage and speed capability.

The Netgain Warp 9 is very similar to the ADC motor and intended to be a drop- in replacement. It is rated at 35 ft-lb torque at 220 amps and 18 hp, 2750 RPM at 72 V. These ratings double to 36 HP and 5500 RPM at 144 V, which is the maximum recommended.

Both the ADC motor and the Warp 9 retail for about $1800 and the packaging and freight will add $200.

Axial Field DC

Agni motors produce high quality, high cost axial flux DC motors to the designs of the original inventor, Cedric Lynch. The familiar Etec/Perm/Briggs & Stratton axial flux motors are 8" in diameter by 2 1/2" long with an axial permanent magnet field and radial conductors. They weigh around 25 lb and put out 12 HP, equivalent to an Advanced DC 7" X91-4003 weighing 75 lb. The weight saving is due to the much more compact magnetic circuit of the axial field comprising much less iron and somewhat less copper, along with modern high strength permanent magnets.

The Agni 155R is 27 cm (10 5/8") diameter by 125 mm (5") long weighs 20 kg (44 lb) and puts out the same power as an ADC FB1-4001 or a Warp 9 which are 16" long and weigh 170 lb.

AC Variable Frequency

The pioneers of high performance, variable speed AC motors driven by variable frequency solid sate Power Electronics Units (PEU) for electric vehicles are AC Propulsion of San Dimas, CA. They started with the design of the Sunraycer for a solar vehicle competition across Australia, moved on to design the drive for the Chevrolet Impact and the GM EV-1 of the late '90s, and now provide quality drives for the likes of the BMW Mini-E. The Tesla roadster drive is very similar. AC Propulsion are also the pioneers of the V2G concept with demonstrations and reports dating back to the very beginning of the 21st Century. They use the combination of the PEU components and the AC motor windings to provide 20 kW of charging or 20 kW of AC inverter output. This offsets the very high cost of AC Propulsion drives of $25-30,000 (if you can get them at all).

Azure Dynamics provided variable speed AC vehicle drives at somewhat lower cost, but with less sophistication, much larger, heavier motors and no V2G capability. They have since folded.

An economical AC drive is now available from High Performance Electric Vehicle Systems, formerly High Performance Golf Carts, and carried by a number of distributors. They offer a range of motors at 48, 72 and 96 Volts designated AC-20, AC-30 and AC-50 with matching controllers from Curtis. Only the AC-50 with matching Curtis 1238-7601 controller has the continuous power required for truck conversions. It is a very attractive option, featuring built in regenerative braking and higher speed capability (6500 rpm) than DC motors. It will probably be longer lived with less maintenance due to the absence of brushes. It is a drop-in replacement for the other NEMA C face motors. The down side is higher first cost of approximately $4000 per motor because each motor needs a dedicated controller.

Chains & Sprockets

Carbon fiber-reinforced chain drives such as Poly Chain from Gates Rubber Co, have taken over the motorcycle final drive chain market due to their extended life, no required lubrication and minimal noise. They are the only high power speed reduction system that can operate in the open air. This simplifies the current application greatly because there is no need to enclose the drive shaft in an oil-tight casing. Eight mm pitch Poly Chain has the flexibility and speed capability for this application and chains 21 mm wide have a continuous power rating giving a safety factor of almost 2:1

Goodyear NRG (goodyearep.com) belts are very similar except that the sprockets have herringbone teeth and are self centering. This is good for keeping the belts from chafing on the sprocket flanges, but requires a double driven sprocket construction as in **Drawing 24**.

Programmable Logic Controllers

A millennium 3 CD12 PLC from Crouzet available through Newark Electronics for $218 has been used successfully, but failed in service. The Eaton ELC-PC12NNDR-1 is similar and there are a number of other manufacturers. You will need a programming cable at $116 to program the PLC to do what you want.

Instruments

Analog instruments are helpful in monitoring the performance of your conversion. A Volt meter to monitor the battery capacity and an ammeter to measure current and thus power are a minimum. Westech instruments to do this are available from EVA and other distributors. Lighted instruments are essential as you will be driving at night. There are a number of combination instruments specifically for EVs to measure voltage, current and amp hours.

Materials

The most convenient source for materials such as aluminum, steel, plastic, etc. is McMaster Carr. As with tools, fastenings and other industrial supplies, they have everything at reasonable prices and quick delivery. The various parts to be specially machined as well as other purchased components in the Material Specification **Drawing 30** are keyed to McMaster Carr catalog numbers.

Engineering Data

Mechanical

Basic units:

The SI system

This system of units, sometimes called the metric system, is standard in all countries except the United States. It was supposed to become the standard here starting forty years ago, but inertia and the size of the U.S. market have kept familiar engineering units in use here till the present. The automotive industry, however, is strictly metric. SI is convenient in that everything is decimal and there is minimal need for conversion constants.

The standard SI unit of mass is the kilogram (kg).
The standard SI unit of length is the meter (m).
The standard SI unit of time is the second (s).
The standard SI unit of force is the Newton, (N), (units named after persons are capitalized).
The acceleration of gravity at the earth's surface is 9.78 m/s.
One kg weighs 9.78 Newtons at the earth's surface, (one sixth as much on the moon).
A force of one N will accelerate 1 kg at 1 m/s.
The standard SI unit of electric potential is the Volt (V).
The standard SI unit of electric charge is the Coulomb (C).
The standard SI unit of electric current is the Ampere (A), 1A=1C/s.
The standard SI unit of power is the Watt, (W), 1.0 A at 1.0 V=1.0 W.
Also a mechanical force of 1N at a velocity of 1m/s exerts a power of 1W
The standard SI unit of energy is the Joule (J)
A power of 1W, mechanical or electrical, exerted for 1s equals an energy of 1 J.
The standard SI unit of torque is the Newton-meter.
A torque of 1 N-m at 1 RPM equals 0.1047 Watts.
Multiples of units are thousands, kilo-, millions, mega-, billions, giga-
Fractions of units are thousandths, mili-, millionths, micro-, billionths, nano-.

The Engineering, English or US Customary system
The standard engineering unit of mass is the slug, (s), (believe it or not!).
A more common unit of mass is the pound-mass (lbm) equal to 1/32 slugs.
The standard engineering unit of length is the inch, ("), (in.),
    or the foot, ('), (ft), equal to 12 inches.
The US measure of long distances is the mile, 5280 ft.
The standard engineering unit of time is the second, s.
The standard engineering unit of force is the pound-force (lb, or more properly lbf).
The acceleration of gravity at the earth's surface is 32 ft/s.
One slug weighs 32 lbf at the earth's surface.
One lbm weighs one lbf at the earth's surface.
A force of 1 lbf will accelerate one slug at 1 ft/s^2 or one lbm at 32 ft/s^2.
The engineering electrical units V, C, A, W are the same as SI.
The engineering mechanical power unit is the horsepower HP=550 ft-lbf/s.
The equivalence between mechanical and electrical power is 745.6 Watts per HP.
The equivalence between Engineering and SI torque is 0.738 lbf-ft per N-m.
A torque of one ft-lbf at 1 RPM is a power of 0.0001990 HP.
The standard Engineering unit of energy is the British Thermal Unit or BTU.

One BTU = 778 lbf-ft = 1055 J = 252 Calories (an obsolete metric unit of energy). Multiples of units are the same as for SI.

Sample Calculations

Batteries

Cells are rated by two numbers, Voltage and capacity in Amp-hours. The Voltage rating is the average potential at which the cell can deliver the total charge expressed as current times time or Ampere-hours, One (Ah) = 3600 ampere-seconds = 3600 Coulombs. The charging voltage will be higher than the rated voltage, and the difference between the average charging voltage and the rated voltage is the efficiency of the cell, typically 85-95%.

Cells can be combined into batteries by connecting them in series and in parallel.

A number of cells, n, connected in <u>series</u> have the same capacity as a single cell in Amp hours, but a Voltage that is equal to n times V.

A number of cells, n, connected in <u>parallel</u> have the same voltage as a single cell, but a capacity equal to n times C Amp hours.

The <u>energy</u> content of a battery pack is equal to the energy in each cell V x Ah (Watt hours) times the number of cells in the pack, however connected.

The <u>power</u> capacity of the pack is pack voltage times current.

Current is often expressed as a fraction of the pack capacity. A 100 Amp current from a 100 Ah rated cell is 1C. A 200 Amp current is 2C, which is around the safe limit for continual draw from large format lithium-ion batteries, although 8-10 C is possible for 10 seconds.

Chargers and Charging Standards

Chargers are rated by voltage and current or power. Clearly the charger has to operate at a voltage slightly higher than the battery rating to drive charge back into it.
Electric Vehicle chargers are segregated by SAE J-1772 standard levels of input voltage and power.

Level 1 chargers operate on 120 V AC and cannot take more than 16 Amps from a 20 Amp circuits, so they are limited to 120 x 16 = 1920 Watts, or just under 2 kW.

Level 2 chargers operate on 240 V single phase AC and can draw up to 80 Amps, or 19,200W, (19.2 kW).

There will be 200-450 V DC quick chargers also separated into level 1, 80 A and level 2, 200 A, i.e. up to 90 kW!

There is a J-1772 "Combo" specification which combines Level 2 AC and Level 1 DC quick charge connections.

In addition to the SAE standard there is the CHAdeMO standard used in Japan for quick charging offering u p to 62.5 kW of high Voltage DC.

Controllers

Controllers are rated for DC voltage and maximum current. Typically the continuous current is something like 1/3 the maximum rating. For example the popular Curtis 1231 controller is rated at a maximum of 144 V and 500 Amps, 72 kW, (96.6 HP), for two minutes. The one hour rating is 225 A or 32.4kW, (43.5 HP).

AC controllers are rated for Voltage and power. For example the AC Propulsion controller takes 336-360 VDC input and can deliver 150 kW maximum or 50 kW continuously.

Motors

Motors are rated by voltage, maximum current and power, continuous current and power and insulation temperature rating. The maximum motor voltage must be the same or greater than the battery pack voltage to avoid overloading it. Although electric motors can withstand short term overloads in current and power they cannot safely withstand overloads in voltage and speed.

The limiting power is determined by the motor's ability to dissipate heat and keep the wire insulation from frying. It pays to be conservative in design to avoid expensive problems. A motor like the ADC FB1-4001 or the Warp 9 is rated at 144 V with Class H insulation which can go to 180°C. The power level at which the motor reaches that temperature vary with current, speed and ambient temperature.

Torque and temperature are primarily determined by the current through the motor, and many DC motors will specify a torque per Amp ratio.

DC motor speed is limited by voltage and many motors provide an RPM per volt ratio. The higher the voltage and the faster the motor runs, the more power it delivers at a given current and temperature, limited by brush wear and the ability of the commutator to stay together at high speed.

DC motors typically provide a performance map for a given voltage with torque or Amps as the abscissa (X-axis) and speed, power and efficiency as parameters.

AC motors typically provide a map with speed on the abscissa and torque and power as parameters. Typically the torque is constant, limited by maximum current from zero RPM up to some maximum power point, and then drops off linearly. This leaves power roughly constant at the maximum power capability of the controller up to the maximum speed.

Poly Chains, Sprockets

Gates Rubber, the supplier of Poly Chain provide a very comprehensive catalog allowing engineering calculations of power and speed capability of various Poly Chain drives. One starts with the maximum speed, wheel diameter and rear axle ratio of the vehicle in question. This gives the drive shaft speed.

For example, the Gen 11 Ford 150 at the speed limit 65 MPH (65 x 5280/3600= 95.33 ft/s) with a tire radius of 14 inches, (circumference, $2 \pi 14$=87.96 inches or 87.96/12=7.33 ft) has a wheel speed of 95.33/7.33=13.00 s^{-1} (revs per second). The rear axle ratio is 3.55, so the drive shaft speed is 13.00x3.55=46.15 s^{-1} or 2769 RPM. Since the motor limiting speed for 9" motors is 5500 RPM, a two-to-one drive ratio is selected. For the Agni with a speed limit of 3600, a ratio of 1.30:1 or less is required. The 67/53 tooth ratio chosen for the Agni drive gives a ratio of 1.26.

The Gates catalog has a table of horsepower ratings for various width and pitch Poly Chains vs. RPM of the faster shaft and diameter of the smaller sprocket (which is on the faster shaft). Eight mm pitch, 21 mm wide chain can deliver 61.3 HP at 5500 RPM.

Since the Warp 9 motor with a 204 V battery pack can only produce 102V x 180 A x 90% efficiency = 16.52 kW or 22.18 HP continuous, and 102 x 500 x .9 = 45.9 kW (61.6 HP) absolute max power as limited by the controller, this is adequate.

The tension in the belt at 61.3 HP and 5500 RPM with a 40 tooth sprocket having a pitch diameter of 4.01 inches (101.8 mm) is 61.3 x 550 = 33,715 lb ft/ sec divided by the belt speed 3.1416 x 4.01/12 x5500/60 = 96.29 ft/sec or 350 lbf! They are pretty strong! Also they have to be pretty tight to avoid slip.

Bibliography

Here are a couple of my related articles and a few books I have found helpful.

Experience with a Compound Lead-Acid-Lithium-ion Battery, Paul H. Kydd, Current Events, **40**(3&4), p. 6, 18-21, March/April, 2008.

Conversion of a Ford F-150 to a Supplemental PHEV, Paul H. Kydd, Current Events **43**(9) 30-34 September, 2011.

Build Your Own Electric Vehicle, Bob Brant, McGraw-Hill Tab Books, NY, 1994, (Get the original version, not the recent reissue with another author).
The Electric Car, Michael H. Westbrook, The Institution of Electrical Engineers/ SAE Warrendale, PA,

V2G-101, Leonard J. Beck, 2009, Available through the Electric Auto Association, Aptos, CA.

Handbook of Batteries, David Linden and Thomas B. Reddy eds., McGraw Hill, NY, (2002).

References

[i] Bottled Lightning, Seth Fletcher, Hill & Wang, NY, (2011)
[ii] Kydd, Paul H., US Patent 7,681,676, March 23, 2010
[iii] Kydd, Paul H., US Patent Application 2010-0004807, January 7, 2010
[iv] Kydd, Paul H., US Patent 8,360,184, Jan. 29, 2013
[v] Kydd, Paul H., US Patent Application 2012-0152644, June 21, 2012
[vi] Kydd, Paul H., Current Events **46**(1) January, 2014, p. 1 and 10-11
[vii] Handbook of Batteries, David Linden and Thomas B. Reddy eds., McGraw Hill, NY (2002) p 6.10
[viii] Handbook of Batteries, David Linden and Thomas B. Reddy eds., McGraw Hill, NY (2002), p.34.1, 34.5. Special shipping and handling procedures are required for lithium batteries per CFR 49. Toronto Globe & Mail August 15, 1989, "Cellular Phone Recall May Cause Setback for Moli"
[ix] Handbook of Batteries, David Linden and Thomas B. Reddy eds., McGraw Hill, NY (2002) p.34.10
[x] Handbook of Batteries, David Linden and Thomas B. Reddy eds., McGraw Hill, NY (2002) p. 35.1 and "Studies of Li-Ion Battery Market-Year 1999" Yanno Research Institute Ltd. Of Japan (2000) and MacArthur, D.; Blomgren, G.; Powers, R.; "Lithium and Lithium-Ion Batteries 2000", Power Associates (2000).
[xi] N.Y. Times, Sept 19, 2003, p. F-1

[xii] Fellner, J.P; Loeber, G.J.; Vukson, S.P.; Riepenhoff, C.A.; J. Power Sources 119-121 (2003) 911-913

[xiii] Belt, J.R.; Chinh, D.H.; Motloch, C.G.; Miller, T.J.; Duong, T.Q.; J. Power Sources 123 (2003) 241-246

[xiv] Build Your Own Electric Vehicle, R.Brant, TAB Books Div. of McGraw-Hill NY 1994 p232.

[xv] Trojan Battery Co. data sheet.

[xvi] Balanced Belt or Chain Drive for Electric Hybrid Vehicle Conversion, Paul H. Kydd, U.S. Patent Application 2009/0000836 Jan. 1, 2009

[xvii] V2G 101 by Leonard J. Beck, available through Electric Auto Association, San Francisco, CA or from the author at len.beck@V2G-101.com, 2009

[xviii] Kydd, P. H. 2014, "Vehicle-Solar-Grid Integration Test at GridSTAR" Current Events **46**(1), p. 1, 10-11

[xix] This project has been supported by the National Science foundation under SBIR Grant No. IIP-1314675.

[xx] Charged Electric Vehicles Magazine Oct/Nov, 2012 p. 22

[xxi] Charged Electric Vehicles Magazine Oct/Nov, 2012 p. 2

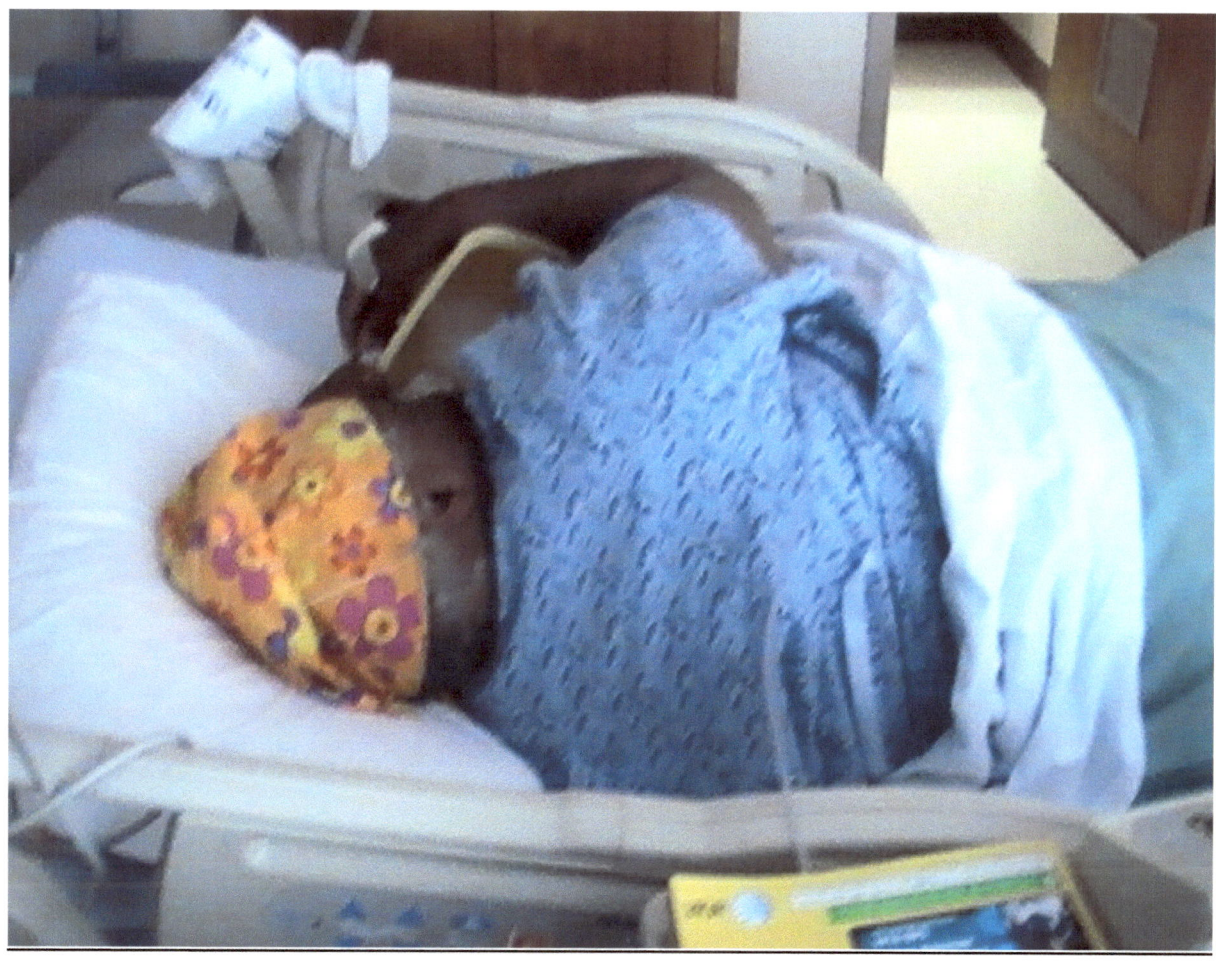

<u>On Thursday October 18, 2012</u>

I had to beg the nurse to give my mother fluids and IV.

Because my mother was still dehydrated, nausea, vomiting, couldn't breathe; at home my mother uses Oxygen o2 dependency with portable tanks and units.

The nurse refuse to keep my mother on oxygen, the entire time my mother was in the hospital at fort sanders regional center in Knoxville TN.

I had to keep pushing the nurse call button for someone to come help my poor weak mother with IV, oxygen, plus fluids and something for her pain.

Before I left the hospital to go home, I told my mother, don't just lay her and suffer, you keep pushing the nurses call button for help.

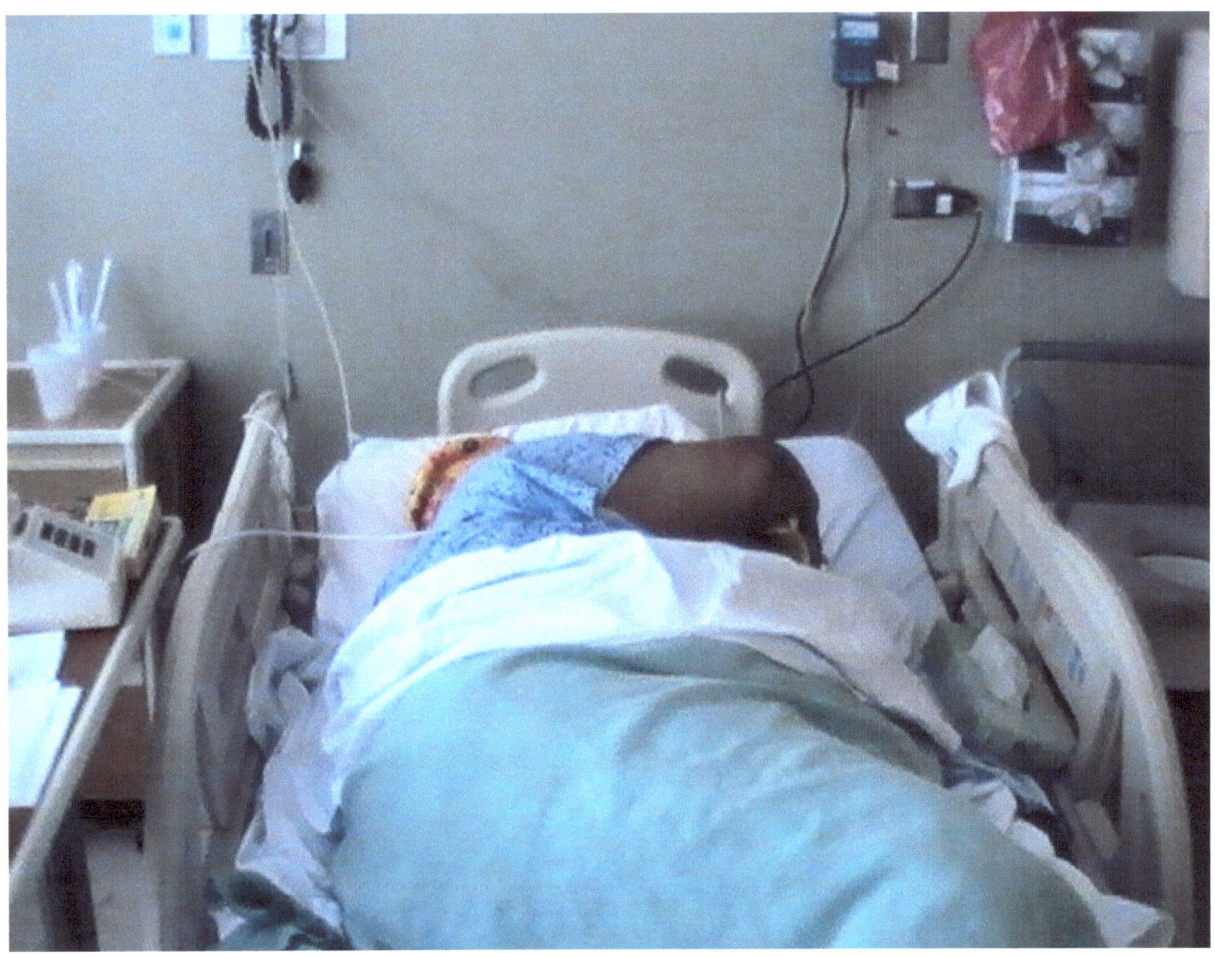

<u>On October 19, 2012 Friday</u>

The nurse called me at 7am and said my mother was being rush for test and MRI. My mother had fallen out the bed; hit her head and injury her right arm trying to reach the nurse call button for help. One of the nurses had move the call button from my mother's bed, when I left to go home the day before. One of the nurses removes the nurse call button, to keep my mother from asking for help. This was how my weak mother got injured at the hospital, trying to reach for the call button that was move. I arrived at the hospital around 4pm. My mother said her body ache all over from the fall. She asks me about any news, about her Medicare approval from social security so I could move her to a better place for help.

I believe my mother suffer with internal bleeding, from the fall.

I'm tired of hospitals getting away with murder, through wrongful deaths.

My mother didn't want me to leave her bed side, she was afraid she was going to die at the hospital, trying to get better. I told her I was still working on trying to get her to the cancer center o f America. I told her I love her and I will see her tomorrow.

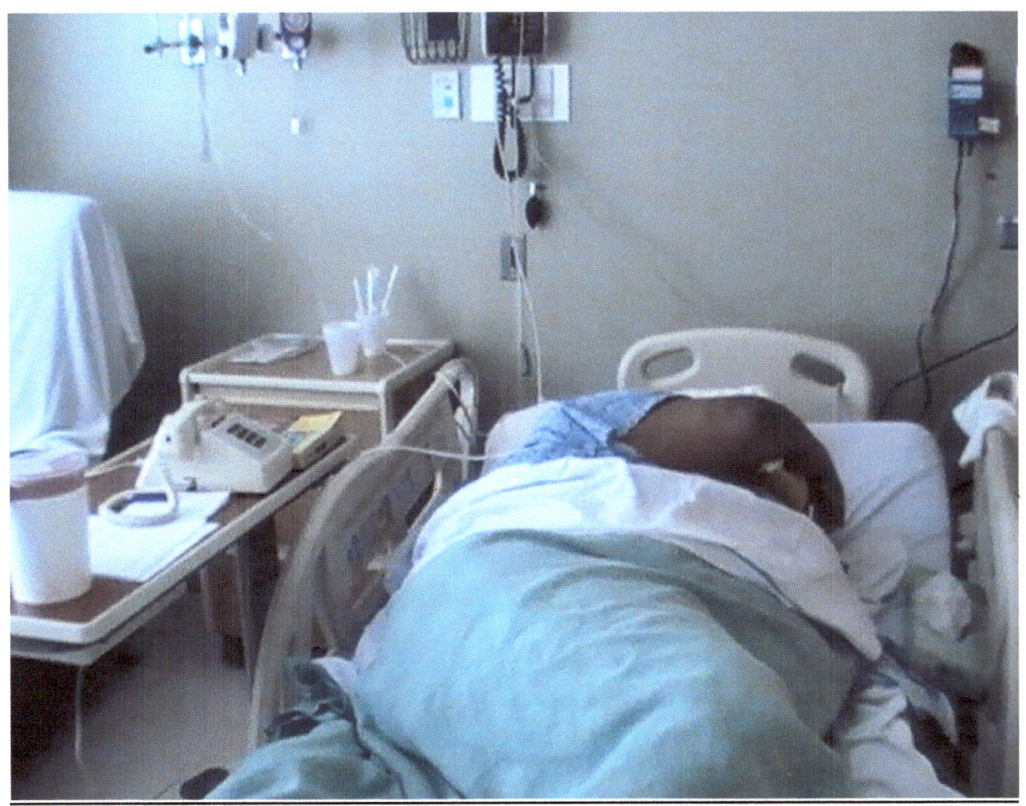

On Tuesday October 23, 2012 my two sons and I went to go visit my mother at fort sanders regional center in Knoxville. My mother had been moved to the 8th floor oncology. She was now in room 814, when I walked into the room, I notice that the doctors had installed a port for chemo in my mother's chest. My mother as so weak tired and yet still dehydrated. I put Chap Stick on her lips because her mouth was very dry and her lips were severely chapped. She was having hard time breathing; she was not on any type of oxygen or IV for fluids. She asks me did she receive approval from social security administration for her Medicare, so I could take her to cancer center of America in Atlanta. I told her that they had denied her Medicare approval, tears just rolled down her face. I told her that I had contact blue cross and shield of Tennessee for insurance for her, so she could go to cancer center of America. Because she had pre existence health problems like cancer, they didn't want to offer her insurance. She said I can't believe no one wants to help me live. Please try another insurance company or just take me home. She didn't want to stay at fort sanders hospital; because she felt that the nurses and doctors weren't doing everything they could to help her. I told her I was going to try another insurance company in Knoxville and if no luck I will drive her to the cancer center in Atlanta at north side hospital tomorrow on October 24, 2012. She said okay, I'll just ride to north side hospital in Atlanta for great hospital service and cancer treatments. I told her I will be back around lunch time, she was so happy to hear, that me and my two sons was taking her to a better hospital in a different state, to help her fight her uterine cancer. I gave her a tight hug and a kiss. And said I LOVE you and we are leaving tomorrow. She looked at me and her two grand children and just smile. She said of all the people in the world, you are the only person

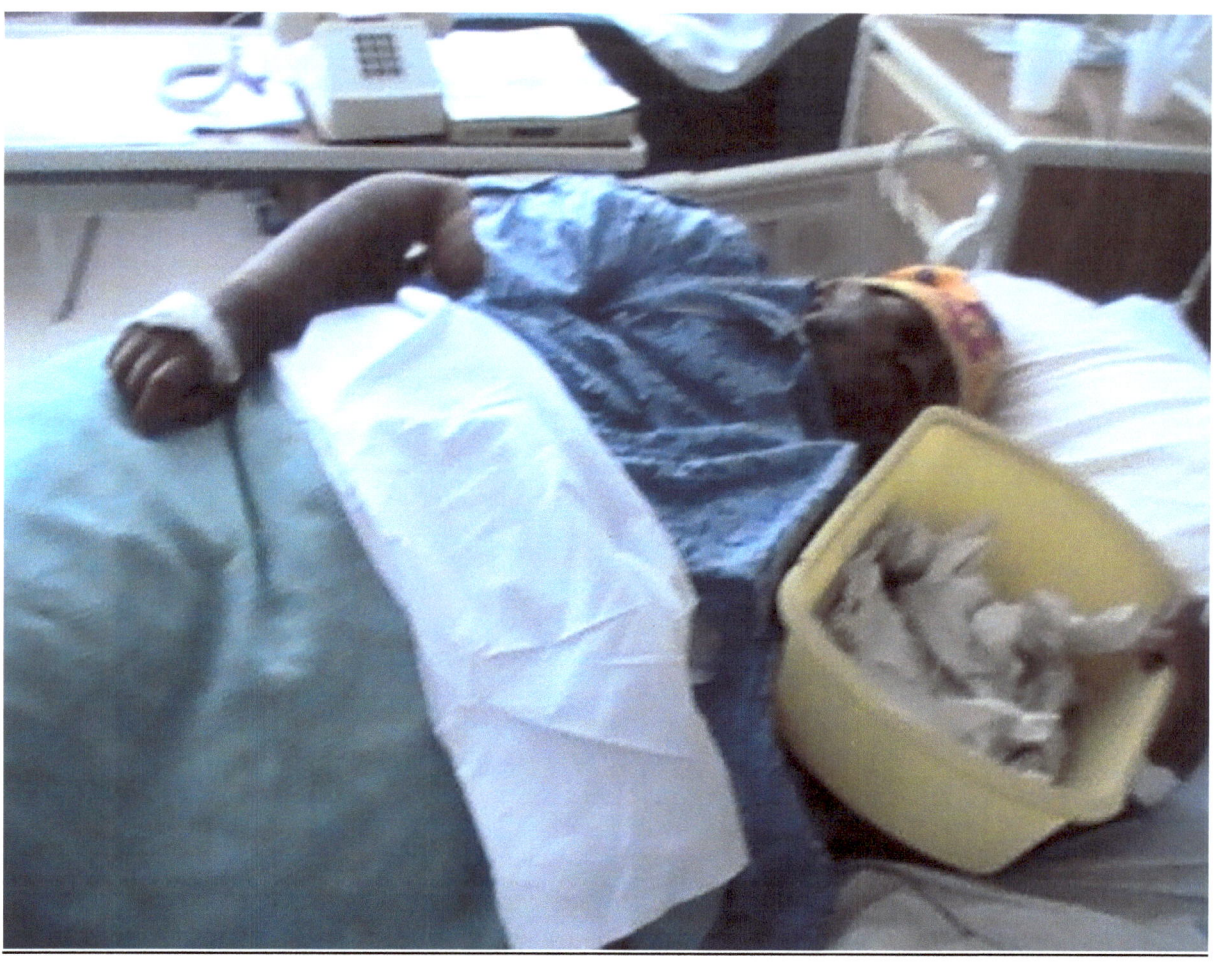

On Wednesday October 24, 2012 at was packing some off my things and moms stuff, to take her to north side hospital in Atlanta Georgia.

The nurse from fort sanders regional medical center in Knoxville, she called me at 9am.

She said that one of the nurses had given my mother a stronger dose of morphine and another nurse again gave her the same pain medication without checking the last time my mother received pain medication.

This pain overdose of morphine within minutes from each other, cause my mother to be unconscious and not responding at all. The nurse said they had to move my mother to the 2nd floor ICU overflow section 5.

I arrived at the hospital around 4:30 on the bus after I pick up my son from school.

When I arrived at the ICU department, the nurse said my mother was not allowed in more visitors for the day. I told her I need to see my mother. She told me to come back tomorrow; the nurse in ICU was really rude. She got loud and was very short to me. She was by far the worst nurse I ever met in my life. She had the worst attitude ever; I knew I was going to move my mother to a hospital in Atlanta the next day. Since the nurse was just full of hate and very evil.

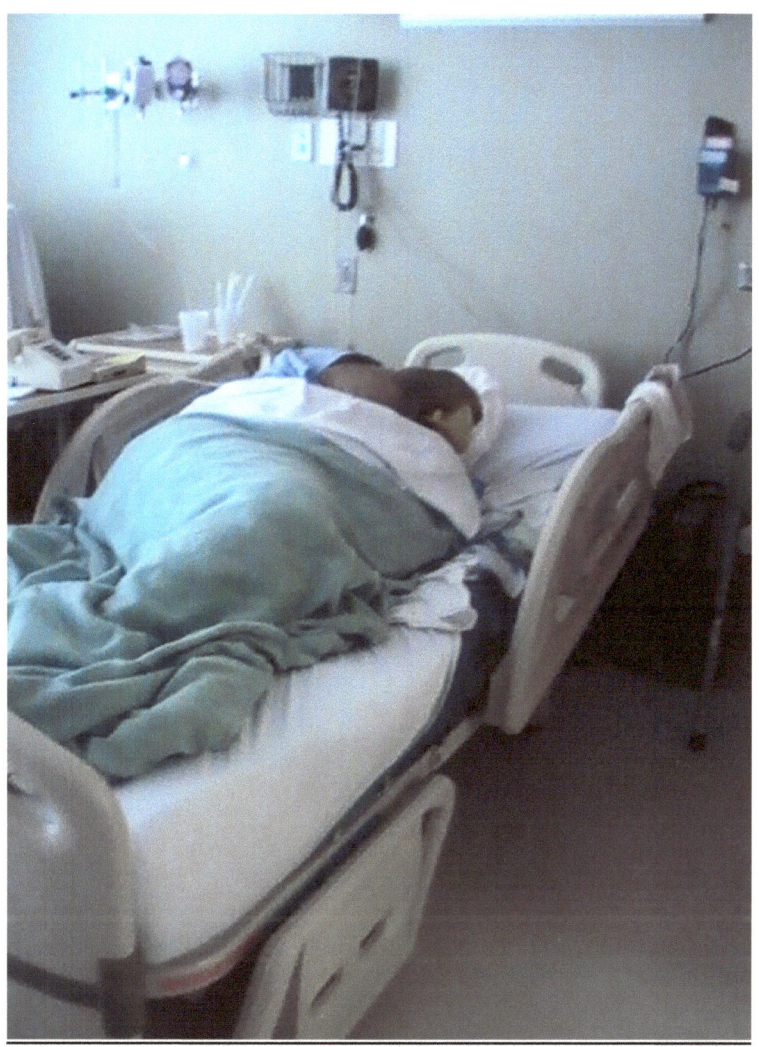

On Thursday October 25, 2012.

The nurse working in ICU said they had to put my mother on life support yesterday after I left the hospital. She said that all of my mother's organs had stop working, and she was still unconscious and not responding at all since she was overdose on Wednesday. She said the only thing was left for me to do was to, remove her off life support. I told her I will be there in an hour, my heart just burst with tears, the pain I was feeling on my way to the hospital was worst than given birth. I was heartbroken on what had happen to my mother; my heart was just shatter in to tiny pieces. All I kept saying was, lord help me. I can't lose my mother, I have not tried everything. I needed just a little more time. When I arrived at the hospital, I walked into the ICU floor where my mother was at. When I enter the room and saw my mother, I was sick. My mother looked like she had been injected with lethal injection. The hospital looks like they had done lethal injection of euthanasia to facilitate death to my mother who was in a lot of pain from her cancer. My mother face was swollen; both of her hands had been strapped down to the bed. My

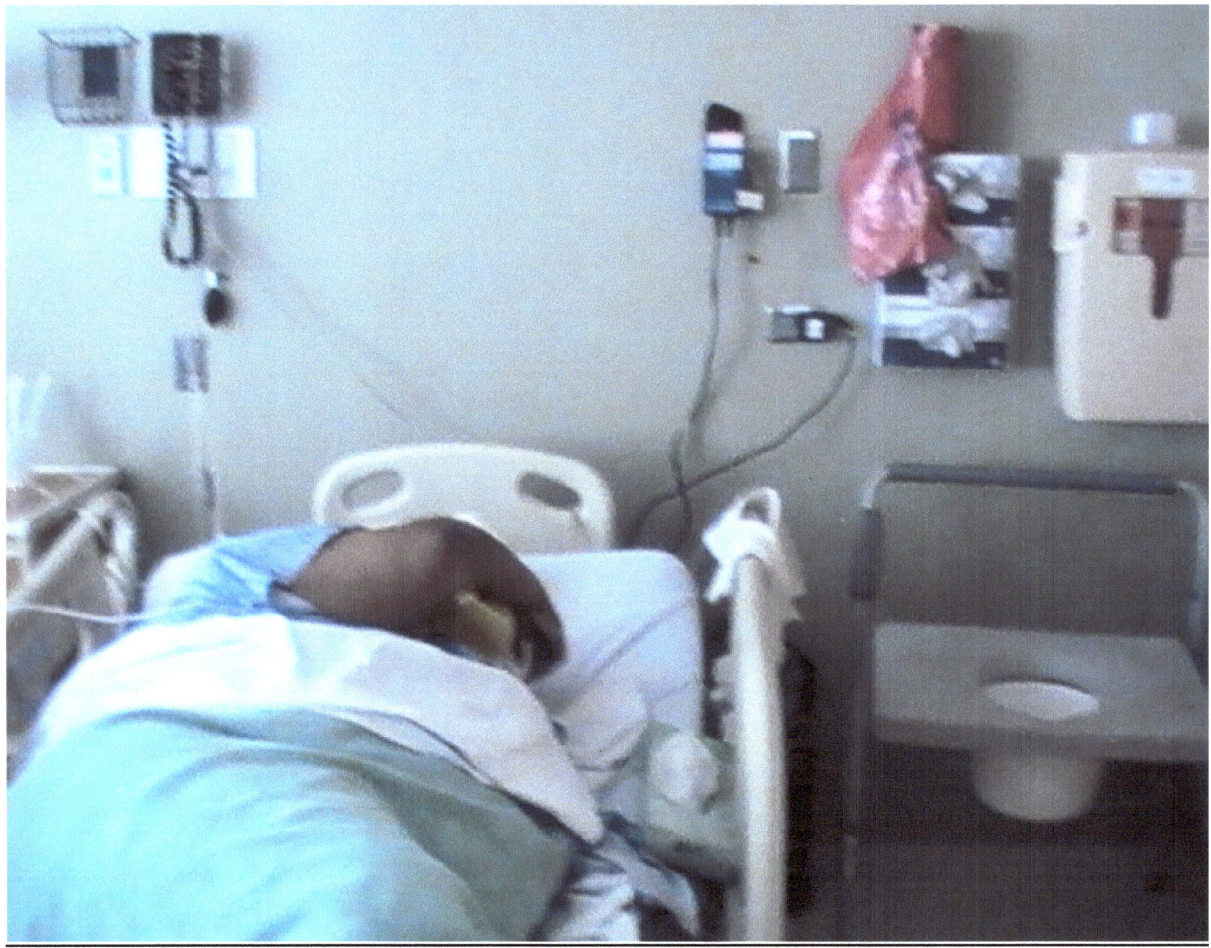

On Friday October 26, 2012

I removed my mother off life support around 12:20am.My mother died from an overdose of pain medications monitor by the doctors of fort sander regional medical center.

I was very sad my mother had to endure unfair hospital and healthcare treatments.

Rose Mortuary on broad way of Knoxville TN came and pick up my mother's body and hour later. She remained with them until October 31, 2012 for crimination services. Today I'm filing a wrongful death law suit against the hospital in my mother's death.

I'm my heart, I believe my mother didn't die at fort sanders regional hospital, my mother was murder.

If my two sons get sick today or tomorrow. I would have to take them to a hospital outside of the state of Tennessee, otherwise they would receive the same heath care and treatments my mother received for years. And she is now dead.

I would like to see justice in my mother's death and everyone who played a part in not saving her life.

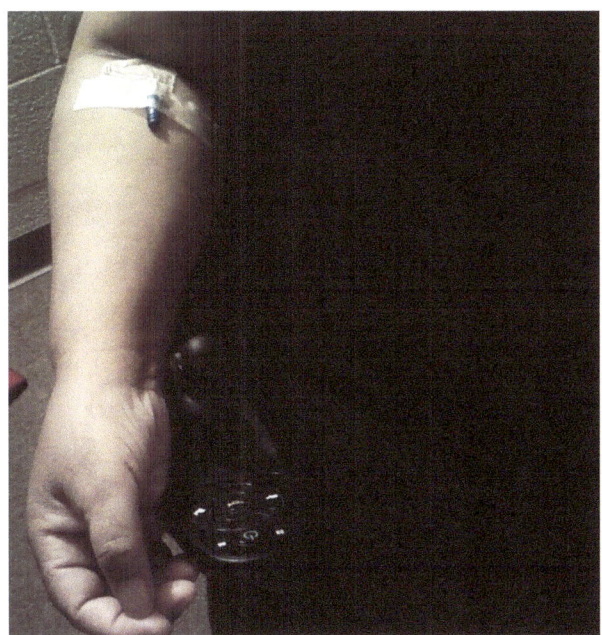

I always tell my mother, when she is having chest pains to pull the cord in her bathroom for help, since I don't have a car to take her to the emergency room.

When my mom, called and told me about the needle still in her arm.

I told her to call the hospital, because the bus for me to come see her doesn't run until 6am.

She says she phoned the hospital right away to ask what she should do.

The emergency room receptionist at Fort Sanders Regional Medical Center said, my mom needed to get a ride back to the emergency room, to remove the needle.

My mom told her she was brought there by ambulance and hospital sent her home in a taxi.

And her daughter don't have a car, she rides the bus to come visit her.

The emergency room said they were too busy to send someone to my mother to remove the needle, and that she needed to find a ride or catch a taxi.

This was around 4am in the morning; my mother knew I didn't have a car, or money for a taxi from west Knoxville to east Knoxville, then to Fort Sanders Regional Medical Center.

I caught the first bus going to my mothers, house after two bus transfers it took me almost two hours on the bus one way.

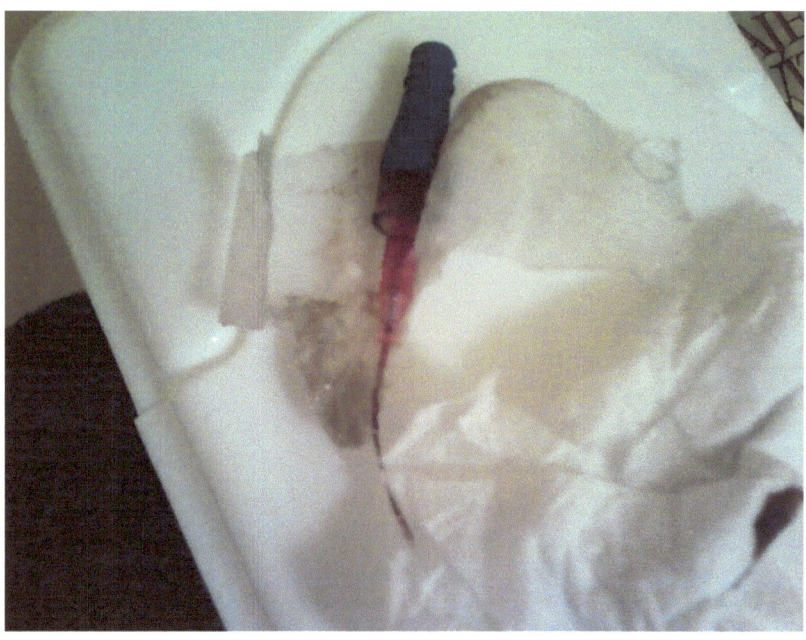

She was still having chest pains, when the hospital sent her home, without any medication to control her chest pains at home.

My mom was treated at Fort Sanders Regional medical center in Knoxville TN, on April 17, 2011.

And four hours later, she was discharged around 3am in the morning.

She was still weak after recovering from chest pains;

I can't believe my mom was sent home from hospital by staff nurse who didn't care to remove an intravenous needle from my mother's arm.

The needle was full of blood, her bruised arm covered in tape, when Theresita Fields was discharged from Fort Sanders Regional Medical Center emergency department by a nurse.

My mother called me around 4 am in the morning; she was still having chest pains.

She was in the bathroom, and notice the need still in her arm, the ambulance took her to the hospital.

When my mother notices the needle in her arm, she panic and took another nitro stat **tablet** for chest pains.

My mother is disable and don't drive, I'm her daughter who checks on her twice a week, when the bus is running.

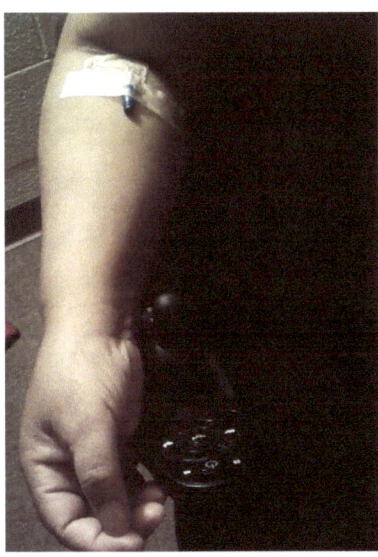

On Sunday April 17 20011 my mother Theresita Fields was having chest pains, and stomach pains (peptic ulcer, she was also very nausea, dehydrated and still vomiting. she pulled the emergency cord in her bedroom for help. The ambulance transported my wheel chair bound mother, to Fort sanders regional hospital in Knoxville .My mother was still having chest pains, when the hospital sent her home, without any medication to control her chest pains at home. My mother was only in the ER at Fort sanders regional hospital for 4 hours, the hospital discharge my mother around 3 am in the morning. My mother was still weak after recovering from chest pains and stomach pains (peptic ulcer, she was also very nausea, dehydrated and still vomiting. I can't believe my mother was sent home from the hospital, by the nurse, who was too busy to remove the intravenous needle from my mother's arm. The needle was full of blood; her bruise arm was covered in tape, when Theresita Fields was discharge from the hospital.

The hospital staff wouldn't allow the ambulance to transport my wheel chair bound back home, so they called a cab for my mother at 3am. The cab drivers drop my sick mother off in front of the building.

My mother wheel chair was in the house. My mom had to sit outside of the building, with the intravenous needle still in her arm. She was waiting for someone to help her get inside the building.

And to help her sit in her wheelchair and her cell phone.

My mother was outside sick and in pain; she had no wheel chair or her Oxygen o2 dependency for her COPD disease.

My mother called me around 5am in the morning; she told me how the hospital treated her.

And how much pain she was still in, my mother refused to return back the hospital that morning, after how she was badly treated .My mother went to the bathroom and notice the needle still in her arm, when she saw the needle, she panic and took another nitro stat tablet for her chest pain, to prevent a heart attack. My mother is disabling, and wheel chair bound, I'm her daughter Dionne Fields who takes care of her and visit her at least four times a week, with my two sons.

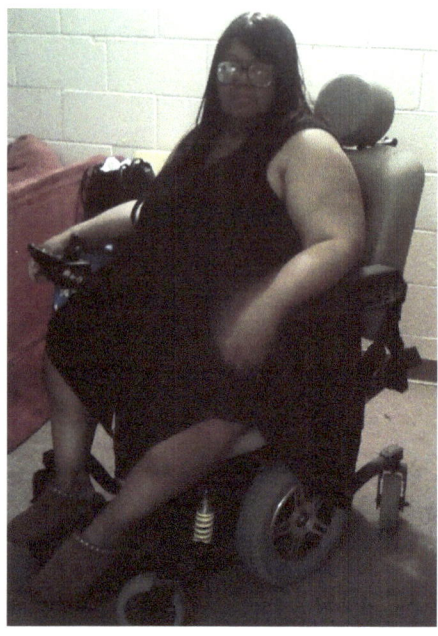

When I arrived at my motherhouse around 8 am, she was still in the bathroom sitting in her wheel chair.

There was blood everywhere; my mom had tried to remove the needle herself.

She didn't want me to see her with the needle still in her arm, when the hospital refused to help her.

I just wanted to get her in her own bed, and cleaned up her bathroom.

This was a nightmare for me as a daughter, who loves her mom very much.

My mother, could have bleed to death, she could have died.

She was already having heart problems and chest pains.

I believe if my mother were a white woman, this would have never happen.

On Monday January 3, 2011- Monday January 10, 2011
My mother was hospitalized and diagnose with peptic ulcer disease at Fort Sanders Regional Hospital. The doctor said her gastric ulcers, were benign. And it was no need to do, further exams for her acute abdominal pains and medical management her gastric ulcers.

Chapter Two

The Untold Story

This page you would see copies of legal documents and how one in the state of Tennessee, that tried to help my mother.

They did nothing to help her or to save her life.

The day my mother died, I lost my best friend and my mother all in the same day.

To help heal my broken heart, I'm telling the true story about my mother's untold story.

I want to help stop hospital negligence and wrongful deaths that happens every day in hospital.

My mother has had physician's Mutual insurance for years, her insurance wouldn't award me any money to bury her for all the years, she has paid

Physicians Mutual Insurance Company®
Physicians Life Insurance Company®
members of the Physicians Mutual® family

Physicians
Mutual®

2600 Dodge Street
Omaha, NE 68131-2671
www.PhysiciansMutual.com

Claim Services:
PO BOX 2018
Omaha, NE 68103-2018

Customer Service:
PO BOX 3313
Omaha, NE 68103-0313

Theresita E Fields
1837 Linden Ave Apt 166F
Knoxville, TN 37917-8037

October 8, 2012
9191

000157

POLICY NO.085-619-268
Theresita E Fields

Dear Theresita E Fields,

We know how easy it can be to overlook sending a payment in the day-to-day rush of things...

...so we thought it was important to write you to let you know that your Physicians Life insurance policy lapsed a short time ago. That's the bad news.

But here's some very good news! If you send your premium within 10 days, we guarantee reinstatement of your insurance policy.

That's all there is to it. As soon as the payment of $141.00 is received, your insurance policy will be paid to November 8, 2012.

Why not put your payment in the mail today--while it's fresh in your mind.

Sincerely,

Kathy Anderson

Kathy Anderson
Assistant Vice President
Customer Service

Policy Information Card - Beneficiary

12/24/2009

Insured Name:
Theresita Fields
Policy Type: Whole life Policy
Current Benefit Coverage: 10,000
Additional Benefit Coverage: 0

Beneficiary's Name:
Dionne L. Fields
Daughter

N228-14

Radiation Treatment DR. Daniel D. Scrapenoth   Thompson Cancer Survival Center

Name: _Theresita Fields_____   8/22/12

Date of Exam: _____   Time: _____

Exam: ___Vaginal Cylinders :_____

Where: _____#1 Monday. August 27, 2012 8:30 AM

Notes: _____#2 Thursday August 30, 2012 9:00 AM

_____#3 Tuesday September 4. 2012 9:00 AM

_____

_____

October 26, 2012

✳ My Mother Theresita Fields, was never told
by her Radiologist that her cancer had spreaded
when she had 3 exams Back in August and
September 2012 at Thompson cancer survival center.

Visitation times are set to make sure
that the patient has time for needed
procedures, meals, and tests. We ask
that you please respect the visitation
times and guidelines listed below

Visiting Hours

9:15	–	10:00	am
1:00	–	2:00	pm
4:30	–	5:00	pm
9:00	–	10:00	pm

The Untold Story. **Tenn care of Tennesee** didn't murder my mother; they just <u>denied</u> her all the medical necessity for her to live.

State of Tennessee
Department of Finance and Administration
Bureau of TennCare
310 Great Circle Road
Nashville, Tennessee 37243

Theresita Fields
1837 Linden Ave. Apt.# 166F
Knoxville, TN 37917

8/16/2012

RE: Applicant: Theresita Fields
Date of Birth: 10/16/1948
Effective Date:
PAE Control Number: 2012215-267152

Dear Theresita Fields:

On 08/02/2012 12:00:00 AM, you applied to get home care through the TennCare CHOICES Program in a **new** home care group (CHOICES Group 3) that's for people at risk of going into a nursing home. This home care is to help delay or prevent your need for nursing home care.

To get this kind of home care through CHOICES, you must not be able to do one of these things by yourself (without help from someone else) most days of the week:

- Transfers
- Walking or using a wheelchair
- Eating
- Toileting
- Knowing where you are and who people are (called orientation)
- Expressing wants and needs, and understanding and following simple instructions
- Taking medicine
- (If you have dementia) behaviors like undressing or wandering that place you at risk.

We reviewed your Pre Admission Evaluation (PAE) application. **Your PAE application was denied.**

This means that you don't meet the medical (level of care) requirements to get home care through CHOICES Group 3.

This is based on Tenncare Rule 1200-13-01-.10(4).

We looked at your PAE and the medical records we got with it. We looked at your functional assessment and

History & Physical

SSI didn't murder my mother; just <u>denied</u> her medicare to attend cancer center of america in Atlanta.

M3

Social Security Administration
Retirement, Survivors, and Disability Insurance
Important Information

Southeastern PSC
Birmingham Social Security Center
1200 Rev. Abraham Woods, Jr. Blvd.
Birmingham, Alabama 35285-0001
Date: October 3, 2012
Claim Number: 223-68-0968 A

0000372 CTPML3 1A   0.440
THERESITA FIELDS
APT 166F
1837 LINDEN AVE
KNOXVILLE TN   37917-8037

Information About Medicare

You recently stated that you should have Medicare in November
2012 based on Social Security disability benefits. A review of
your Social Security record shows that you are not drawing
disability benefits. You are drawing retirement benefits.
You, therefore, are not due Medicare until you obtain age 65.

If You Have Any Questions

We invite you to visit our website at www.socialsecurity.gov on
the Internet to find general information about Social Security.
If you have any specific questions, you may call us toll-free
at 1-800-772-1213, or call your local Social Security office at
1-866-331-8636. We can answer most questions over the phone.
If you are deaf or hard of hearing, you may call our TTY
number, 1-800-325-0778. You can also write or visit any Social
Security office. The office that serves your area is located
at:

                    SOCIAL SECURITY
                    8530 KINGSTON PIKE
                    KNOXVILLE,TN 37919

                    SEE NEXT PAGE

My mother has receive disablility for years & SSI ,would not approve her medicare stage 4 cancer.

SUPPLEMENTAL SECURITY INCOME
NOTICE OF RECONSIDERATION - DISABILITY

From:
   Social Security Administration

JAN 1 9 2004

DATE:
   THERESITA E FIELDS   Claim Number: 223-68-0968
   2101 WALTON WAY
   APT 302
   AUGUSTA, GA 30904

You must meet certain medical and non-medical requirements to be entitled to disability benefits. We have found that you meet the medical requirements for disability benefits.

WHO MADE THE DECISION

The determination on your claim was made by an agency of the State. It was not made by your own doctor or by other people or agencies writing reports about you. However, any evidence they gave us was used in making this determination. Doctors and other people in the State agency who are trained in disability evaluation reviewed the evidence and made the determination based on Social Security law and regulations. In making our decision we took into consideration the opinion of treating and/or examining medical sources.

HOW WE MADE THE DECISION

The following reports were used to decide this claim in addition to those listed on our previous notice.

DR CARLOS C TAN report received 11/19/2003

Additional reports were not obtainable; however, the one(s) shown had enough information to evaluate the condition.

You said that you became disabled on April 4, 1987 because of hypertension, back and knee pain, breathing problems, chest pains and kidney stones. However, the medical evidence does not show that your condition was severe enough at that time to meet our requirements. In medically evaluating your condition, both the severity and type of impairment were reviewed. We also took into consideration vocational factors such as your age, education and work

223-68-0968   FIELDS,THERESITA E

The Untold Story.I have filed dozens of complainst for years,for **hospital negeligence** concerning my mother, and nother was ever done.

U.S. Department of Justice

Civil Rights Division

Disability Rights Section - NYA
950 Pennsylvania Avenue N.W.
Washington, DC 20530

Dionne Fields
3807 Middlebrook Pike Apt. 108
Knoxville, TN 37921

JUL 1 9 2011

  Re: Fort Sanders Regional Medical Center
  Date Received by DOJ: June 09, 2011
  DOJ Number: 202-70-0 Theresita Fields

Dear Ms. Fields:

  The Disability Rights Section of the Civil Rights Division has received your correspondence.

  The circumstances you describe do not appear to raise an issue that we are able to address. We believe that the agency or organization indicated below is in a better position to assist you. For your convenience, we are returning your correspondence. If you so choose, you can file your complaint with:

    Ms. Anita Giuntoli
    Associate Director
    Office of Quality Monitoring
    The Joint Commission
    One Renaissance Boulevard
    Oakbrook Terrace, Illinois 60181
    (630) 792-5867
    1-800-   4-   )
    (www.jointcomm

    Sincerely,

    Catherine C. O'Brien

    Catherine C. O'Brien
    Civil Rights Program Specialist
    Disability Rights Section
    Civil Rights Division

374223

The Untold Story.I have filed dozens of complainst for years,for **hospital negeligence** concerning my mother, and nother was ever done.

The Joint Commission

July 28, 2011

Dionne Fields
3807 Middlebrook Pike - Apt. 108
Knoxville, TN 37921

Regarding: Fort Sanders Regional Medical Center
Incident #81495CZX-11618MOC   Theresita Fields

Dear Ms. Fields:

Thank you for bringing your concerns about the above named organization to the attention of The Joint Commission.

As the leading evaluator and accreditor of health care organizations in this country, The Joint Commission takes seriously any information received about one of our accredited organizations. As such, we will review your submission for possible non-compliance with accreditation requirements, and determine what action should be pursued. The possibilities are:

- Asking the organization for a written response
- Conducting an immediate for-cause survey at the organization to evaluate the issues
- Evaluation of your concerns as part of a regular survey at the organization
- Including your information in a database for future consideration or
- Determining that the issues underlying your complaint are not related to The Joint Commission standards, and therefore, cannot be evaluated by The Joint Commission.

Thank you for taking the time to inform us of this issue. If you need to contact us, please feel free to do so and refer to the above incident number.

Sincerely,

Office of Quality Monitoring

Headquarters
One Renaissance Boulevard
Oakbrook Terrace, IL 60181
630 792 5000 Voice

The Untold Story.I have filed dozens of complainst for years,for **hospital negeligence** concerning my mother, and nother was ever done.

DEPARTMENT OF HEALTH & HUMAN SERVICES   **OFFICE OF THE SECRETARY**

Voice- (404) 562-7886, (800) 368-1019   **Office for Civil Rights, Region IV**
TDD- (404) 562-7884, (800) 537-7697   **61 Forsyth Street, S. W.**
(FAX) - (404) 562-7881   **Atlanta Federal Center, Suite 3B70**
http://www.hhs.gov/ocr/   **Atlanta, GA 30303-8909**

                                    July 28, 2011

Ms. Dionne Fields
3807 Middlebrook Pike Apt. 108
Knoxville, TN 37921

Our Transaction number: 04-11-130076-CP-CR
                    Theresita Fields

Dear Ms. Fields:

Thank you for your correspondence received on 05/05/2011 by the Department of Health and Human Services, Office for Civil Rights (OCR). _letter about mom ._

We are in the process of reviewing your correspondence to decide whether OCR has authority and is able to take action with respect to the matters you have raised. We will complete our initial review as quickly as possible.

If you have any questions, please contact:

Office for Civil Rights, Region IV
61 Forsyth Street, S.W.
Atlanta Federal Center, Suite 3B70
Atlanta, GA 30303
1-800-368-1019

When contacting this office, please remember to include the transaction number that we have given your file. That number is located in the upper left-hand corner of this letter.

                        Sincerely,

                        _by_ Roosevelt Freeman
                        Regional Manager
                        Region IV

The Untold Story.I have filed dozens of complainst for years,for **hospital negeligence** concerning my mother, and nother was ever done.

TENNESSEE DEPARTMENT OF HEALTH
BUREAU OF HEALTH LICENSURE AND REGULATION
DIVISION OF HEALTH CARE FACILITIES
227 FRENCH LANDING, SUITE 501
HERITAGE PLACE METROCENTER
NASHVILLE, TN 37243
TELEPHONE (615) 741-7221
FAX 615-741-7051
www.tennessee.gov/health

June 16, 2011

Dionne Fields
3807 Middlebrook Pike
Apt 108
Knoxville, TN 37921

Dear Ms. Fields,

We received a complaint letter from you regarding your mother's (Theresita Fields) visit to Fort Sanders Regional Medical Center on 04/17/2011.

I would like to speak to you regarding your concerns.

I am available Monday – Friday 8:00am – 3:30pm at 1-877-287-0010.

Sincerely,

Anne Sanders, RN

The Untold Story.**Blue care of Tennesee** didn't murder my mother; they just <u>denied</u> her all the medical necessity for her to live.

STATE OF TENNESSEE
BUREAU OF TENNCARE
TennCare Solutions Unit
P.O. Box 000593
Nashville, Tennessee 37202-0593

Theresita Fields
1837 Linden Ave Apt 166F
Knoxville, TN 37917

Matter ID: 12-10-027-558367 (This is the number we use to track your appeal.)

(10/16/12) mom's Birthday

Dear Ms. Fields:

We received your health plan change request. You said you needed your health plan changed because you need medical treatment.

✗ Your health plan change request has been denied, but you still need help with your care.

We want to make sure that you get the care that you need. TennCare has contacted your current health plan, BlueCare.

BlueCare will contact you to help you get the care that you need.

If you don't hear from BlueCare in 5 days from when you get this letter, call TennCare Solutions at **1-800-878-3192** to let us know. Have this letter with you when you call.

Do you need help with this letter? Is it because you have a health, mental health, or learning problem or a disability? Or, do you need help in another language? If so, you have a right to get help, and we can help you. Call TennCare Solutions at **1-800-878-3192**.

- Do you have a **mental illness and need help with this letter?**
  The TennCare Partners Advocacy Line can help you.
  Call them for free at **1-800-758-1638**.

- If you have a hearing or speech problem you can call us on a **TTY/TDD** machine.
  Our TTY/TDD number is **1-866-771-7043**.

The Untold Story.**Dr. Fred Merkel Promary care doctor** didn't murder my mother; he just refuses to fix her broken wheel, so she could continue her daily radiation treatments.

09/20/2012

THERESITA FIELDS
1837 LINDEN AVE APT 166F
KNOXVILLE, TN 37917

Dear THERESITA,

Due to the fact that you have not been seen by a provider at Cherokee Health Systems for a long time, you w
need to call 544-0406 to make an appointment for evaluation regarding your request to repair your wheelchair.

Sincerely,

Provider: Fred Alan Merkel DO 09/20/2012 11:13 AM

Document generated by: Darly Drown, LPN 09/20/2012 11:13 AM

T

The End

I love you darling

Theresita Fields

In Loving Memory.

October 16, 1948- October 26, 2012